Organizational Risk Management and Sustainability: A Practical Step-by-Step Guide

Organizational Risk Management and Sustainability: A Practical Step-by-Step Guide

by
Robert Pojasek

CRC Press
Taylor & Francis Group
Boca Raton London New York

CRC Press is an imprint of the
Taylor & Francis Group, an **informa** business

CRC Press
Taylor & Francis Group
6000 Broken Sound Parkway NW, Suite 300
Boca Raton, FL 33487-2742

© 2017 by Taylor & Francis Group, LLC
CRC Press is an imprint of Taylor & Francis Group, an Informa business

No claim to original U.S. Government works

Printed on acid-free paper

International Standard Book Number-13: 978-1-4987-2466-1 (Hardback)

Visit the Taylor & Francis Web site at
http://www.taylorandfrancis.com

and the CRC Press Web site at
http://www.crcpress.com

This book is dedicated to the memory of my uncle

and first mentor, Walter J. Pojasek.

Contents

Section I   Foundation for Risk Management and Organizational Sustainability

Section II   Structure for Planning and Implementing
Organizational Sustainability

Section III   Monitoring, Measuring, and Improving Organizational Sustainability

Preface

One cannot understand sustainability without consideration of the *perspective* for examining this topic. It is possible to look at sustainability from the perspective of an organization, a corporation, or a community (including city, state or province, or country). There are also a number of other perspectives people use when discussing sustainability: the needs of future generations, planet Earth (e.g., climate change and water scarcity), other stakeholders without a voice (e.g., plants and animals), and the urgency of single causes (e.g., eliminating the use of fossil fuels). The diversity of perspectives on the topic of sustainability explains the confusion over the meaning and value of sustainability. What makes matters worse is when people mix these perspectives in a single program without knowing that perspective is so important to understanding and success in the application of sustainability. This book has been written from the perspective of an *organization*. I hope it will be your guide for making some sense out of this confusion.

Organizations are social entities and comprise the basic building block of society. All the major standard-setting organizations have recognized the importance of defining their terms from the perspective of an organization. Standards of Australia started their work on risk management in 1990. This led to the publication of the risk management standard, AS 4360, in 1995. Standards of Australia advanced the need to embed risk management and how organizations were managed. This standard became an international standard, ISO 31000, in 2009. It provides the means for embedding risk management and the assessment of uncertainty (i.e., opportunities and threats) into what every organization member or employee does every day.

To provide a proper foundation for what is presented in this book, information from these standards was compared with published information on organizational development and theory; people, culture, and principles; decision making, sense making, and knowledge management; risk management; and sustainability. All this information was focused on organizations and the embedding of programs into how the organizations are operated, rather than the practice of having separate initiatives for sustainability programs. This information is provided in the first section of the book: "Foundation for Risk Management and Organizational Sustainability."

To ensure effective processes, efficient operations, and efficacious strategy within every organization, the plan–do–check–act structure of the harmonized high-level structure of the International Organization for Standardization was used. Section II examines the planning and implementation components of the International Organization for Standardization standards, with the focus on use by any kind of organization or any size. Section III is focused on monitoring and measurement of the organizational sustainability program, along with continual improvement, innovation, and learning. The closing chapter presents the highlights of the British Standards guidelines on resilient organizations.

In the appendix, a case is presented on the "virtual" Olive Hotel located in the context of Cambridge, Massachusetts. This case was used in my online course at Harvard University. It is included to provide some ideas on how to use the materials in the book with the organizations that you are seeking to improve. This was the concept that helps fulfill the need for a practical step-by-step guide for the people that read this book and wish to embed sustainability in one or more of the many originations to which they belong.

What developed in the work to prepare this book was a realization that organizational sustainability is the new unambiguous sustainability. It is now possible to articulate unambiguous details of a sustainability program once the organization and its context are known. Living in a time where social media is proclaiming the "death of sustainability" every day, it is comforting to know that organizational sustainability is alive and well—even though many people have yet to discover this approach. If we could help each of our organizations to become more sustainable and convince suppliers, customers, civic groups, and government organizations to become more sustainable, we will be much more likely to have some tangible influence on how this will affect future generations and the overall health of all the systems (environmental, social, and economic) at play here on this planet.

Of course, this will not happen overnight. However, the effectiveness should help us progress at a rate greater than that experienced since the 1992 Earth Summit. This book represents a change in how we approach sustainability. Without reconstructing the fundamental information supporting a change in how we view sustainability, this book would not have been written. There will be additional changes once it is realized that there is a uniform perspective for having conversations and action with this new unambiguous sustainability for organizations.

Robert B. Pojasek, PhD
Cambridge, Massachusetts

Acknowledgments

I want to acknowledge my colleagues and clients who have influenced my thinking, consulting, writing, and teaching in the area of organizational sustainability. You can find their names on my LinkedIn contacts list and the members of my LinkedIn group—the Sustainability Working Group. I would also like to thank all those who have participated in my course at the Harvard Extension School and many other short courses and consulting assignments around the world.

I wish to personally thank Holly Duckworth and Andrea Hoffmeier for providing me with the inspiration to complete the work on this book and introducing me to Taylor & Francis publishers. The CRC Press team was terrific to work with as the ideas and thoughts were captured and presented in this book.

I would especially like to thank a former student, Cheri Mohr, for her help providing editorial advice for many of the chapters and for assembling the Olive Hotel case presented in the book's appendices. Her efforts were noteworthy in helping this book earn its subtitle—"A Practical Step-by-Step Guide." Thank you, Cheri.

I would also like to acknowledge the many organizations that write standards. They include the International Organization for Standardization, European Foundation for Quality Management, British Standards Institute, Baldrige Performance Excellence, Standards of Australia and New Zealand, and SAI Global (a former employer and manager of the Australian Business Excellence Model). All these standard setters have adopted the perspective of an organization in their standards. The development of the harmonized high-level structure by the International Organization for Standardization has made it possible for organizations to integrate and put to use the entire spectrum of standards that are currently available and to include those currently under development. These standards are at the heart of what is presented in this book. I hope you will consider purchasing the standards that are referenced in this book. While I was able to provide you with an idea of "what" needs to be implemented, their standards and guides contain the best practices that will help you understand "how" to best use their fine work in your organization.

Finally, I would like to thank the other authors of the works cited in this book. I have acknowledged their works with endnotes and references. Please read their articles and books to provide to supplement your knowledge of risk management and organizational sustainability. I am especially grateful to Olivier Serrat (Asian Development Bank) for his contributions to the literature that made many of the concepts presented in this book possible.

What started out as a course for understanding sustainability at the organization level has become an important perspective on how all sustainability programs should be operated, maintained, and evaluated. The effects of uncertainty in the world have made it imperative that organizational sustainability becomes the basic building block of the organizations that constitute corporations, cities, regions, and countries. I would like to thank those of you that are interested and able to take this work to the next step, so we can realize what we have struggled with since the work leading up to the 1992 Earth Summit in Rio de Janeiro.

Author

Robert B. Pojasek had just about completed his undergraduate degree (Rutgers University, New Brunswick, New Jersey) in chemistry when he found himself participating in the first Earth Day. This event led him to an interdisciplinary PhD degree (University of Massachusetts, Amherst) researching the transformation and transport of manganese in the Quabbin Reservoir in Massachusetts and a consulting career in the field of process improvement. He became intrigued with the concept of sustainable development during the lead-up to the 1992 Earth Summit held in Rio de Janeiro, Brazil.

Dr. Pojasek started consulting in the nascent field of sustainability and spent 14 years teaching an online course entitled "Fundamentals of Organizational Sustainability." More than 1500 students from 50 countries around the world helped shape his perspective on this important topic. The experiences from his consulting and his teaching provide the basis for this book. With the adoption of harmonized management systems at the international level, many corporations are beginning to realize that each of their facilities is different because of its unique context. Dr. Pojasek is still providing risk management and sustainability consulting (i.e., Pojasek & Associates LLC) to organizations of all sizes. The urgency of his work has increased as corporate sustainability programs have struggled with integrating sustainability into what every employee does every day---both within the company and the other elements of the organizations value chain. His work has been expanding into urban centers that are promoting sustainability as a means of sustainable development.

Dr. Pojasek has served as the chair of the board of the Corporate Responsibility Association and a board member of the International Society for Sustainability Professionals. He continues to teach an online course on organizational sustainability at Harvard University and has been working with other universities interested in adopting the perspective of organizational sustainability.

Section I

Foundation for Risk Management and Organizational Sustainability

The first six chapters of this book provide a foundation of credible published information to support each element of the plan–do–check–act (PDCA) approach to organizational sustainability. It is important to support the development of organizational sustainability programs with established disciplines such as organizational development, decision making, sense making, knowledge management, systems of management, and risk management. The relationships to established disciplines will enable organizational sustainability to move beyond an "initiative-based" practice that is added onto an organization, rather than being built into that organization. Other established disciplines will be added over time as the practice of organizational sustainability continues to develop in organizations. This foundation information will be referenced in the other sections of this book. Putting the information up front may help the reader find more applications for this information or lead the way to introduce yet other topics.

The second section of this book presents the planning and implementation (i.e., the "do" phase of the PDCA cycle) of an organizational sustainability program. There are eight chapters covering the topics necessary in this area.

Chapter 14 provides a summary demonstrating the necessity of interlinkages between the various elements to help improve the sustainability program.

In the third section of this book, the focus is on monitoring and measuring the performance of the organizational sustainability program. The section concludes with the role of continual improvement, innovation, and learning in helping an organization become resilient.

1

Understanding Organizations

Organizations are being increasingly pressured by their customers, the government, and their stakeholders to make a commitment to sustainability with its three responsibilities: environmental stewardship, social well-being, and economic prosperity in the community. The digital revolution is changing the face of sustainability by shifting the focus to organizations because they have the control and sphere of influence to make a difference in how they are structured, managed, and operated to meet these responsibilities. Organizations are collaborating with other organizations to provide innovative products and services that help meet the demands made by the public. Our understanding of sustainability is changing now that it can be viewed from the perspective of an organization.

It becomes confusing when there is a conversation about sustainability and everyone involved presents their views from different perspectives. For example, there are the perspectives of future generations, the planet Earth, "stakeholders" without a voice (i.e., biodiversity), and even a sustainability report itself. To understand the perspective of an organization, those involved in these conversations need to appreciate the concepts that are used to describe and analyze organizations and how organizations affect the way we live our lives.[1]

What Is an Organization?

An organization is a social entity that has objectives (whether explicitly stated or not), a deliberate operating structure, and a coordinated activity system and is linked to its external operating environment.[2] Organizations come in many sizes and kinds: sole proprietorships, microbusinesses, companies, corporations, partnerships, not-for-profits, nongovernmental organizations (NGOs), government departments and agencies, houses of worship, families, and institutions. These organizations can be a single unaffiliated operating unit or operating as part of a large multinational corporation. The term *organization* does not typically apply to government acting in its sovereign role to create and enforce law, exercise judicial authority, and carry out its duty to establish policy in the public interest or honor the international obligations of state.[3]

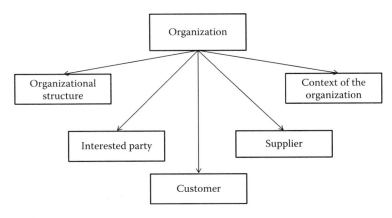

FIGURE 1.1
Relationship diagram for an organization. (Adapted from ISO, Quality management systems—Fundamentals and vocabulary, ISO/DIS 9000, ISO, Geneva, 2014.)

Organizations are made up of people in a structure with an arrangement of responsibilities, authorities, and relationships. Leaders create a structure and commit resources to achieve the organization's purpose—its strategic objectives. An organization interacts with its customers, suppliers, and external stakeholders. All this is true whether the organization is unaffiliated or a part of a larger organization. Every organization has an internal and external context that it operates within. A relationship diagram helps to illustrate the organizational relationships covered by topics in this book[4] (Figure 1.1). Organizations shape our lives in many ways, and each of us belongs to multiple organizations.

Organizational Theory

The field of organizational theory provides us with the tools needed to evaluate and understand how organizations operate.[5] Organizational theory is based on the systematic study of organizations by academic researchers. Instead of presenting a collection of facts, this field is focused on research into organizations and how the people that belong to them and material resources are organized to collectively accomplish a particular purpose. Concepts have been derived from organizations of many different types located in different external operating environments in countries around the world. Organizational theory is a way to analyze organizations more consistently, accurately, and deeply than one could do otherwise.[6]

There are multiple levels of analysis in the field of organizational theory[7]:

- The first level looks at individuals as the basic building block of organizations.

- The second level is the organization itself.
- The third level involves the structural elements that define how the individuals work together to perform group tasks.
- The fourth level involves organizations grouped together into an interorganizational set or community as part of a "parent" organization.

People are important to organizations. Organizational theory provides a macro examination of people in the organization because it analyzes the organization as a whole. Organizational behavior, on the other hand, is a micro approach because it focuses on the individuals within the organizations as the relevant units of analysis.

The point of studying organizations is to enable us to find ways to improve their performance and effectiveness.[8] To be effective, an organization needs clearly articulated overarching objectives and an appropriate strategy for achieving these objectives. Increasing an organization's effectiveness is not always a simple matter. Different people want different things for an organization, and when change is needed, people often establish change initiatives that are not embedded into how the organization operates. Superimposed on this is the demand that an organization constantly adapt to changes in its external operating environment, cope with increasing size and complexity, and create the right kind of culture needed to help it meet its objectives.[9]

Organizational theory classifies an organization's design on a scale ranging from "mechanistic" to "organic." Mechanistic design is characterized by a centralized operating structure, specialized tasks, formal management systems, vertical communication, and a strict hierarchy of authority. On the other end of the scale, organic design is characterized by a decentralized structure, empowered roles, informal systems, horizontal communication, and collaborative teamwork.[10] The complexity of the external operating environment in the world today is causing many organizations to shift to more organic designs. Mechanistic designs are more effective when the change in the external operating environment is changing at a slower pace. Means for analyzing the internal and external context of an organization are addressed in Chapter 8.

Organization's Objectives

Objectives are defined as "results to be achieved."[11] Some people use the terms *goals*, *targets*, and *aims*, and sometimes use them interchangeably with each other and with the term *objective*. It is important to be consistent with terminology when attempting to understand the hierarchy and

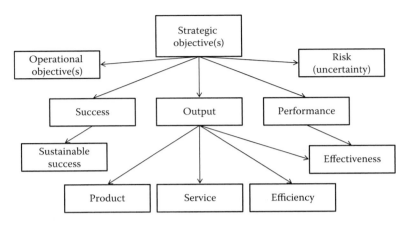

FIGURE 1.2
Relationship diagram for the strategic objectives of an organization. (Adapted from ISO, Quality management systems—Fundamentals and vocabulary, ISO/DIS 9000, ISO, Geneva, 2014.)

organizational components of a top-down, bottom-up process approach by which sustainability is embedded into the day-to-day operations of an organization. Special attention will be paid to carefully defining terms used in this book.

Objectives can be strategic, tactical, or operational. As previously mentioned, strategic objectives relate to the organization's purpose. Tactical objectives are set for each operating unit of the organization, from the top of the organization to the bottom. All the objectives must be linked to the strategic objectives. At the operational level in an organization, objectives can relate to different functions (e.g., financial; environmental, health and safety; assets management; and quality). A relationship diagram of how an organization addresses objectives and other related issues is provided in Figure 1.2.

An organization is deemed to be successful when it meets its objectives. Sustained success involves meeting objectives over a period of time—typically on a 5- to 10-year horizon. To accomplish sustained success, an organization needs to establish a balance between economic and financial interests and those of the social and environmental elements of its external operating environment (discussed in Chapter 8). Sustained success also involves a range of stakeholders of an organization, such as leaders, owners, members, employees, suppliers, bankers, unions, partners, neighbors, the community, and society in a larger sense than the community.[12] Sustained success is possible when organizations consistently address the interests of these stakeholders over the long term.

Organizations, however, operate in an uncertain external environment that may affect the ability to meet their objectives. Risk is defined as the "effects of uncertainty on objectives."[13] Uncertainty is caused by the state of deficiency of information related to understanding or knowledge of an event, incident, or decision that can create "effects." Effects are deviations from what is expected, whether they be positive (opportunities) or negative (threats). These opportunities and threats are generally found by characterizing the internal and external context of the organization. This is presented in Chapter 8. Sometimes people will use the term *risk and opportunity* when discussing risk. However, this is an investment term that substitutes risk for *threat*, one of the two possible effects of uncertainty. Risk assessment focuses on the management of uncertainty involving both opportunities and threats. More information on the management of risk is presented in Chapter 5. The topic of uncertainty is presented in Chapter 11.

Organizations measure the "performance" of their ability to meet strategic objectives in terms of "effectiveness" (i.e., the extent to which planned activities are realized and planned results are achieved).[14] Organizational performance is presented in Chapter 17.

Sphere of Influence

An organization typically has control over the operations found within its internal operating environment. It has some level of influence in its external operating environment. The sphere of influence is defined as the "extent of political, contractual, economic or other relationships through which an organization has the ability to affect the decisions or activities of individuals or organizations outside of its internal context."[15] It should be pointed out that having the ability to influence does not imply a responsibility to exercise that influence.

An organization can exercise its influence with others either to enhance positive impacts on sustainable development or to minimize negative impacts—or both. When assessing an organization's sphere of influence and determining its responsibilities, an organization needs to exercise "due diligence."[16] Different methods for exercising influence may include

- Establishing contractual provisions or incentives
- Public statements by the organization
- Engaging with the community, political leaders, and other external stakeholders
- Making investment decisions
- Sharing knowledge and information with others

- Conducting joint projects with others to address the organization's changing external context
- Understanding the responsible use of the organization's external communications capabilities
- Promoting good operating practices
- Forming partnerships with sector associations, NGOs, and others

The exercise of an organization's influence should always be guided by ethical behavior and other principles and practices of sustainability. When an organization decides to exert its influence, it should encourage other organizations to do the same so that sustainable supply chains can be created for all to use. When the supply chain does not cooperate, alternative actions may need to be considered, including changing the nature of the relationship with these organizations.[17]

Due diligence in a sustainability sense is a proactive process for identifying the actual and potential opportunities and threats associated with environmental, social, and economic impacts of an organization's activities and decisions with the aim of avoiding threats and embracing opportunities.[18] Due diligence also involves the behavior of others where they are found to influence the opportunities and threats in areas where the organization may be involved.

Organizations should not feel compelled to create "initiatives" to demonstrate they are acting sustainably. Rather, they should focus on meeting their responsibly established objectives, complying with laws and regulations, and being capable of managing opportunities, threats, and disruptions encountered in a changing external operating environment. Risk management and sustainability must be embedded in how the organization is operated. The importance of this statement will be discussed further in Chapters 5 and 6.

Sustainability of Organizations

Organizations around the world are becoming increasingly aware of the need for, and benefits of, operating in a responsible manner. The object of sustainability is to create a strategic focus on stewardship within an organization that will contribute to sustainable development.[19] Stakeholders are increasingly exhibiting interests in the organization's performance in relation to its impact on the environment, the well-being of the employees and the community, and the shared value that it creates in the community, as well as for its customers.

The perception and reality of an organization's performance with regard to sustainability can influence[20]

- Its competitive advantage
- Its ability to attract and retain members, employees, customers, clients, or users
- The maintenance of morale, commitment, and productivity in the internal context
- The view of investors, owners, donors, sponsors, and the financial community
- Its relationship with companies, governments, the media, suppliers, peers, customers, and the community within which it operates

To achieve sustained success, the organization should adopt an approach to sustainability that is based on the principles of risk management.[21] Additional information on sustainability is presented in Chapters 6 and 14.

Resilient Organizations

There is growing recognition that organizations must be resilient if they are to survive in an uncertain world. Resilience enables an organization to effectively deal with complexity and ambiguity, learn from experience, make investments in leadership and culture, and create stronger internal and external networks. Resilience is neither a set of principles nor a "destination." It is a journey that takes place through the practice of risk management and is informed by organizations, people, knowledge, and technology, as well as interactions among all these.[22]

To make the journey to become a resilient organization, it is important to study organizational theory for available organizational configurations.[23] If people establish and maintain an organization to do things that are not already being done, it is not possible to design a resilient organization. Using today's thinking, we will have only mechanistic organizations, and not organic organizations that can survive in an uncertain world. A resilient organization must create bridges between what is referred to as "organizational silos."[24] This calls for collaboration, coordination, capability, and connection, and is not easy to accomplish within most organizations. A model has been developed to move from functional silos in organizations to "systems"—a necessary transition to enable resilient organizations, the organizations of the future. More on this topic is presented in Chapter 20.

Whether conventional, sustainable, or resilient, it is important to remember that all organizations consist of people and to further understand how people, principles, and culture work within an organization.

Endnotes

1. Daft, 2013.
2. Daft, 2013.
3. ISO, 2010.
4. ISO, 2014.
5. Daft, 2013.
6. Daft, 2013.
7. Daft, 2013.
8. Daft, 2013.
9. Daft, 2013.
10. Daft, 2013.
11. ISO, 2014.
12. ISO, 2014.
13. ISO, 2009.
14. ISO, 2014.
15. ISO, 2010.
16. ISO, 2010.
17. ISO, 2010.
18. ISO, 2010.
19. ISO, 2010.
20. ISO, 2010.
21. ISO, 2010.
22. Serrat, 2013.
23. Serrat, 2012.
24. Serrat, 2010.

2

People, Principles, and Culture

People are the essence of an organization. Their full involvement enables their abilities to be used to help the organization meet its objectives. As the leaders establish the mission statement and strategic objectives of the organization, they should create and maintain an internal environment in which people can have a high level of involvement and engagement.

All of us belong to many different organizations—families, schools, work, civic and professionals groups, houses of worship, and nongovernmental organizations (NGOs). Organizations are social entities that exist when people interact with one another to perform essential activities that help the organization meet its strategic objectives.[1] This chapter moves from the organizational focus to the people that constitute the organization.

Human Resources

Human resources is the term used to describe the collection of people and all the associated structures within which they associate with each other in an organization.[2] From the people perspective, this is called human capital. Human capital represents the physical, intellectual, emotional, and spiritual capacities of any individual.[3] There is also a focus on social capital at a higher level. Social capital represents a focus on people at the community and national levels, rather than the people at the organizational level.

There are four important building blocks of dealing with people in organizations.[4]

Open Communications

Members or employees of an organization need to have access to the information necessary to conduct their activity. Communication between leaders and employees must be clear, timely, and supported by facts.[5] This information should reside in a knowledge management function that can meet the needs of the organization. The knowledge management system is needed to enable decisions to be made at any level in the organization (see Chapter 3). Without information, people cannot fulfill their roles in an organization.

Communication tells members of an organization what the mission statement and the associated strategic objectives mean to them. Mission and objectives must be clearly articulated by the leaders to the members throughout the entire organization.

Ideas and feedback should also be obtained *from* the members or employees. This can only happen when there is "freedom from fear." Members and employees do not want to be exposed to blame, reprisals, and other consequences administered by the leaders when they volunteer information.[6]

Employees can best participate and work toward the strategic objectives when they have knowledge of facts and data. Facts and data are critical to the success of any process improvement activities. This data orientation makes decisions much more impersonal[7] and can minimize perceptions of blame.

Objectives are a top-down initiative in organizations. The goals are set at the bottom and generate the feedback that will establish whether the objectives have been met. When the discretionary efforts of members or employees in this feedback loop are exerted toward improvement so the objectives will be met, this is an example of "ownership" behavior of the employees.[8] Employees need to be fully engaged in meeting the strategic efforts through their contributions with the work they perform. Importantly, they need to know the results of the strategic-level objectives if they are to sustain their focus on the action plans and goals that they control in the pursuit of meeting the overarching objectives of the organization. All communication, whether one way or two way, is part of the knowledge management system that helps people fulfill their roles in the organization.

Trust

Developing trust within an organization is critically important to its success. In order for management to trust the messages from the members or employees, the employees trust the management not to punish the messenger. Leader behaviors that encourage trust include[9]

- *Engagement*: Two-way communication
- *Consistent communication*: Saying the same thing to all listeners
- *Honesty*: Telling the truth, even when it is awkward to do so
- *Fairness*: Maintaining the organization's policies
- *Respect for opinion of others*: Active listening to understand needs, ideas, and concerns
- *Integrity*: Maintaining the company principles, saying what will be done, and doing what was said

Building trust takes consistent actions over a long period of time.

In this way, employees will grow to respect and learn to rely on one another as they collaborate across all processes within the organization. They will then tend to be ethical in their behavior and seek fair resolutions to difficult situations within the organization. Organizations with a high level of trust create employees with confidence in their leader's vision for the future and the strategic objectives of the organization.

Stability of Membership or Employment

Stability is an important trait for an organization in order to maintain existing members or employees and attract new ones. First, it is important that leadership recognizes the perceived threat that if employees work hard to continuously improve the processes and operations, it will lead to less demand for their services. Management must communicate how it is prepared to retrain or find new ways of keeping people engaged when the worker productivity is improved. In some organizations, there may be seasonal or other predictable fluctuations in the work cycle, creating a need to plan for fluctuations in employment so as not to create a false impression of stability. Some organizations plan for fluctuations by seeking temporary employees to help through the busy times or provide some overtime to the existing employees.

Whatever the circumstances, thoughtful planning and communication are necessary to achieve employment stability. It may not be easy, but it is certainly worth the effort.[10]

Performance Appraisal and Coaching

It is important to get this right. Evaluating the performance of members or employees can be very helpful when done correctly, and incredibly destructive when mishandled. Leaders need to understand that the purpose of the employee evaluation is employee development, and then develop the means needed to execute the program effectively. Many organizations are moving away from the old ways of conducting performance reviews and to systems where employee development is the focus.

Effective organizational development first seeks agreement on the job expectations, and then measures employees' performance against the expectation. A development plan should be prepared that provides coaching, feedback, personal development activities, and support.

Organizational Culture

Culture is often seen as the totality of a society's distinctive ideas, beliefs, values, and knowledge. It explains how people interpret their surroundings.

Cultural theory is associated with the study of anthropology and a host of related fields (e.g., political economy, sociology, and communications). The perspective of culture theory includes the needs common to all people, something that is not easy to achieve, but makes it possible to focus on the whole and the parts, on contexts and contents, on values and value systems, and on strategic relationships between countries and people, as well as the natural environment. The focus of culture is on systems rather than on parts of systems.[11] Recognition of culture allows people to make informed choices and decisions regarding the future and enables us to deal with complexity and fragmentation.[12]

Organizational culture is part of the internal context of the organization, and consists of the attitudes, experiences, and values of the people within the organization. This culture is acquired through social learning and determines the way people interact within the organization, as well as with society outside of the organization. All organizations have a culture, whether it forms organically or is strategically created by leaders. In order to create an organizational culture that supports the organization's mission and strategic objectives, the leaders must understand how a culture is formed within organizations. There are a number of attributes that influence an organization's culture (Figure 2.1). By identifying these attributes, organizations can see which can be managed to help implement and sustain a constructive social change within the organization. None of the components can influence an organization's culture on its own, and none can individually deliver the desired improvements.

Organizational culture varies between organizations more than any other corporate construct. At the lowest levels, an organization is seen as having created an effective culture when members or employees respond positively to policies affecting the items that have the most meaning to them.[13,14]

At the highest levels, it is the culture that helps an organization understand what it stands for and where it is trying to go, the changing nature

Customs and norms	Ceremonies and events
Rules and policies	Objectives and measurement
Management behaviors	Rewards and recognition
Learning and development	Communications
Organizational structure	External context

FIGURE 2.1
Components of organizational context. (Modified from Serrat, O., Asking effective questions, Knowledge Solutions, Asian Development Bank, Manila, 2009, http://adb.org/sites/default/files/pub/2009/asking-effective-questions.pdf.)

of the organization's external operating environment, what matters to the organization, and what resources it has to use as it operates and grows. This knowledge can be used to make the organization more resilient and capable of operating more effectively and confidently in our increasingly volatile, uncertain, complex, and ambiguous world. People say that organizational culture determines how things get accomplished in an organization—what happens and what does not happen.[15]

There is a correlation between the orientation of organizational culture and organizational learning. Organizations that are more successful at implementing knowledge management use both operations—and people-oriented strategies—to enable change.[16]

Using people-oriented strategies, culture can be instilled in each new member as a way to think, feel, and behave while participating in the organization. The culture is even important to members of the organization's value chain and the external stakeholders in the community because all stakeholders participate in the organization's culture. By making the components of the culture explicit, the organization will be in a better position to articulate its culture to its stakeholders.

In any organization, there are a number of observable behaviors, stories, and ceremonies that are shared. These common elements of culture reflect the deeper beliefs and values that are present in the minds of the members of that organization. Common cultural elements provide people with a positive sense of organizational identity, which generates within each member a commitment to beliefs and values that are common to that organization. A positive culture can underpin the vision and strategy of the organization. It contributes to the success of the organization by motivating the members to attain the objectives.

A positive culture also helps the organization understand and address how it adapts to an uncertain external operating environment. It plays an important role in creating a climate that enables learning and innovative response to challenges, competitive threats, and new opportunities. A strong culture encourages adaptation and change that enables the organization to enhance its performance by literally energizing and inspiring members around a shared sense of achievement of the objectives and a higher purpose for the future of the organization. A shared culture shapes and guides behavior so that everyone's actions are aligned with the strategic priority of meeting the organization's objectives.[17]

Risk and Sustainability Culture

Since organizations seek to manage the effects of uncertainty so they can meet their strategic objectives, it is important that they develop a risk-aware culture. Often, there is a direct relationship between the risk awareness in organizational culture and organizations' potential to respond to uncertainties and manage opportunities and threats. The same is true with an organization's sustainability strategies. When these strategies are imposed by initiatives that are not consistent with the organization's culture, it is likely that they will be difficult, if not impossible, to implement. Sustainability strategies should be informed by the risk management and sustainability policies of the organization and coordinated with the governance activities that seek to create and maintain effective management of opportunities and threats.

The process of identifying the opportunities and threats through the determination of the internal and external context (Chapter 8), and the assessment of those effects of uncertainty (Chapter 11), requires human involvement and judgment. Each person involved in this process is affected by preconceptions and unconscious bias. Furthermore, people are influenced by their attitudes toward uncertainty, which can range from threat-adverse, through risk neutral, to opportunity seeking. As you can imagine, these attitudes have a profound effect on how they feel and think about opportunities and threats and influence the judgments they make during the uncertainty assessment process.

As organizations develop an active interest in creating a culture of sustainability, one area where there has been substantial emphasis involves the creation of a culturally competent organization.[18] A culturally competent organization brings together knowledge about different groups of people and transforms it into standards, policies, and processes that make everything work. However, the focus has often been on the creation of standards, policies, and processes, rather than linking the effort to the culture. Of course, you need both.

Core Values and Principles

The core values of an organization are the rules that govern ethical standards, and they are applied in relation to individual behavior, as well as interactions with colleagues and clients or customers as individuals execute their responsibilities. Core values often include norms, such as integrity, ethical behavior, innovation, teamwork, superior quality, outstanding service, social responsibility, and being a good citizen in the community. Many organizations develop a statement of core values to emphasize the expectation that the values be reflected in the conduct of the operations and the behavior of their employees and leaders. Even in an ever-changing world, the core values within an organization are often considered to be constant. Core values are the practices used every day in everything that is done in the organization. Workplace values work best when they are aligned with the personal values people set outside of work.

Large organizations use values to shape their culture in numerous ways. Organizational values can be used to instill a sense of identity and purpose in organizations; add spirit to the workplace; align and unify people; promote employee ownership; attract newcomers; create consistency; simplify decision making; energize endeavors; raise efficiency; strengthen client trust, loyalty, and forgiveness for mistakes; build resilience to shocks; and contribute to society at large. These organizational values can backfire, however, when leaders or employees fail to live up to the underlying norms. Organizational values can also be a source of tension if they conflict with the organizational objectives, often expressed in financial terms, or the underlying organizational culture from which such values are supposed to originate.[19]

Some organizations have engaged in values-driven management improvement efforts, including values training, appraising leaders and employees on their adherence to organizational values, and employing organizational development specialists to help them understand how their organizational values affect the overall performance. The fundamental findings of a study of these practices are as follows[20]:

- Ethical behavior is a critical component of a company's social license to operate.
- Most organizations believe that values influence two important strategic areas—relationships and reputations—but do not see the direct link to growth.
- Most large organizations are not measuring the financial return on their values-driven activities.
- Financial leaders are approaching values more comprehensively.
- Values practices vary significantly by country and region.
- The tone of the top leaders matters.

An interesting organizational values "start-up kit" has been created by the Asian Development Bank.[21]

Guiding principles can be seen as guidelines that drive individual's behavior or mindset as they execute the strategic and operational plans that lead to an organization's success. Risk management and social responsibility principles address the risk-aware mindsets needed for the successful development and subsequent execution of strategic and operational plans. Guiding principles express how an individual in an organization is expected to execute established ethics.

It is important to note that the terms *principles* and *ethics* are sometimes used interchangeably. Some organizations state principles from the perspective of the organization itself and not the individual. Using the organization's perspective provides a way to let outside stakeholders know what the organization stands for. In large organizations, the sustainability or social responsibility program often states principles that represent what the organization as a whole stands for.

Successful organizations have implicitly followed their organizational principles for generations. It is an important part of transparency for the organization to explicitly state its principles so that stakeholders, internal and external to the organization, can see if the organization is true to its ideals.

Functional practices like risk or quality management establish a set of principles that inform and guide an organization's approach to its system of management. Organizations use the principles in conjunction with the design and continual improvement of the uncertainty analysis process. Unlike the organizational features of the process focus and the steps of the uncertainty analysis process, the principles are not presented as specified actions. Instead, the principles serve as the underlying concepts and drivers. The principles provide guidance to both the way the management framework is structured and how the risk management process is applied. All the characteristics of the principles can be used to diagnostically evaluate the effectiveness of their application of the risk management and sustainability management frameworks.[22]

Principles Specified in International Standards

Principles provided within a system of management are defined as a fundamental basis for decision making or behavior.[23] An international risk management standard lists 11 principles.[24] All processes in a system of management used by an organization (Chapter 4) are informed and guided by these principles. This is called "giving effect" to the principles. Employees and other stakeholders should be able to express what each of the stated principles means to them based on the information that is provided by the organization. These principles also inform the leaders' management and their commitment to risk management and sustainability.

Here is a look at what the principles might look like when used for a sustainability effort:

- Our organization exists to create and protect value for our members, employees, customers, and stakeholders.

- Sustainability is an integral part of how we operate, not a "bolt-on" effort.
- Sustainability is embedded into how all decisions are made every day at all levels in the organization.
- Sustainability is systematic, structured, and timely, thereby helping us to address uncertainty posed by our internal and external operating environments.
- Our leaders' understanding of sustainability is based on the best information from the knowledge of our members or employees and the collective wisdom of our entire value chain and all our external stakeholders.
- Sustainability is tailored to our organization to help us meet our strategic objectives in a transparent and inclusive manner.
- Our organization takes human and cultural factors into account in our sustainability program.
- Our vision of sustainability is dynamic, interactive, and responsive to changes in our internal and external operating environments.
- Our organization uses sustainability to help us innovate so that we can continually improve in our quest to meet our objectives.

The organization must seek to make everyone aware and conversant with each of the principles. Knowing how these words are used to describe the principle helps create understanding of each item. Of course, the significance of the principles will vary depending on the role and responsibility of each person in the organization that uses these statements. The focus of developing awareness of the principles should be on how each principle helps the organization better use them to meet its strategic objectives.

Using Principles in the Risk Management Program

It is possible to develop simple tools that can facilitate the check on how each principle is used in the operation of the risk management and sustainability program (Figure 2.2). It starts with the creation of a list of process elements that are important in the system of management (Figure 2.2). These process elements can be used to create a systematic sampling approach that considers various levels and types of an organization's activities, as well as its formal system of governance, leadership, and support. This does not need to be a comprehensive review across all parts of the organization for each element in order to provide a good check of how the principles are given effect in the operation of the risk management and sustainability programs.

Program element	Extent evident (0–10)	Evidence (Bullet points)	Opportunity for effect
Strategy			
Process			
Procedure			
Policies			
Operations			
Support			
Auditing and review			
Performance evaluations			
Continual improvement			
Scoring	Total score	Performance rating score	Performance rating score

FIGURE 2.2
Tool for evaluating the effect of application of each principle. (Adapted from AS/NZS, Risk Management Guidelines, Companion to AS/NZS ISO 31000:2009, HB 436, SAI Global Press, Sydney, 2013, Table 2.)

The evaluations of how principles are being applied when conducted using this approach should help improve the effectiveness of the processes and the way the risk management and sustainability are applied. In some cases, the diagnostic may indicate the need to significantly improve the operations. Changes to operations can be addressed by creating a "corrective action" or by including them in the action plans that are part of the objectives and goals process[25] (Chapter 7).

It is worth taking some time to see how these principles work so well in helping to embed sustainability into the organization. The information presented in this chapter helps to provide a means for personalizing the sustainability program. When asked in an organization, "Who is in charge of the sustainability program?" the answer should be "Everyone is in charge of sustainability!" It should be easy to see that if these principles are given effect by every member or employee of the organization, then sustainability is an important part of how they do their work every day. It is no longer just about "green team" involvement. Instead, it is about how people create effective processes and efficient operations and about the role of everyone to engage with one another to execute the efficacious strategy.

Endnotes

1. Daft, 2013.
2. Garwood and Hallen, 2010.
3. Porritt, 2006.
4. Garwood and Hallen, 2010.
5. Garwood and Hallen, 2010.
6. Garwood and Hallen, 2010.
7. Garwood and Hallen, 2010.
8. Garwood and Hallen, 2010.
9. Garwood and Hallen, 2010.
10. Garwood and Hallen, 2010.
11. Serrat, 2008.
12. Serrat, 2008.
13. Daft, 2013.
14. Serrat, 2009e.
15. Serrat, 2008.
16. Serrat, 2008.
17. Serrat, 2008.
18. Community Tool Box, n.d.
19. Serrat, 2010a.
20. Serrat, 2010a.
21. Serrat, 2010a.
22. AS/NZS, 2013.
23. ISO, 2010.
24. ISO, 2009.
25. AS/NZS, 2013.

3

Decision Making

Organizations achieve sustainable success in part by improving their ability for all leaders, members, or employees to make decisions. In most cases, poor decisions are more of a threat to organizations than events. The success of any sustainability effort depends on the effective use of decision making, knowledge management, and sense making. However, sustainability professionals cannot help the organization improve its decision making by operating in a remote fashion. When sustainability is embedded within the organization, its number one priority should be to understand the decision-making process and seek to instill the three responsibilities of sustainability and the sustainability principles into this very important process.

Decision Making in Organizations

Leaders in an organization need to understand the effects of uncertainty (opportunities and threats) to create a decision-making process that helps the organization meet its strategic objectives. Most people in an organization are involved in decision making, so the process must be clearly defined and carefully monitored within a sustainability program. When an event contributes to creating a threat for the organization, the uncertainty is heightened and the organization is likely to be on the downside of risk. It could be that the event provides the organization with an opportunity that can be realized to help the organization be on the upside of risk. In these cases, an event is an occurrence or change of a particular set of circumstances.[1] The event could be one or more occurrences with several causes, or it could be something that did not happen.[2] However, it does not take an event to influence the risk. The significant opportunities and threats are addressed with decisions where there is no event involved.

Making decisions is part of the risk and sustainability management processes. Making poor decisions or making no decision can jeopardize the ability of the organization to meet its objectives. Risk is created or altered when decisions are made because there is almost always some uncertainty associated with the decision-making process. Risk is always associated with the strategic objectives. Everyone needs to understand that risk taking is an unavoidable part of all the organization's activities. Risk associated with a decision should be understood at the time the decision is made, not after the decision has already been made. Risk taking must be intentional.[3]

Risk management provides the foundation for informed decision making. It needs to be integrated into activities supporting the achievement of the strategic objectives and the decision-making process[4]:

- Decisions made on the organization's strategic issues should take into consideration any uncertainties (opportunities and threats) associated with the external operating environment, as well as changes in the organization's internal context.
- The organization's innovation process should take into consideration not only the identified uncertainties, but also uncertainties related to the human, social, safety, and environmental aspects of the innovation itself, and be managed according to legal requirements, when applicable.
- Plans for significant financial investment should specify the decision-making milestones at which uncertainty analysis will occur.

The organization's policy on risk management should reflect these and other points that influence the ability to meet the strategic objectives. There needs to be a framework that guides how decisions are made so that the process is applied in an effective and consistent way in all decision making. This requires a clear allocation of accountability, supported by skill development and performance review.[5]

Decision-Making Process

In an ideal situation, decision making requires a complete search of all available alternatives, reliable information about their consequences, and consistent preferences to evaluate the potential outcomes. This never happens in the real world. Every person is limited by his or her cognitive and mental capabilities, the extent of knowledge and information available, and conflicts of interest that are not aligned with the organization's objectives. Organizations usually have a decision-making framework that helps control the decision-making process.[6]

Decisions are made at all levels of the organization. They can be[7]

- *Strategic*: Related to the design of a plan of action to achieve an objective

- *Tactical*: Related to the way different parts of an organization are arranged to deliver the objective
- *Operational*: Related to the way individuals work on a daily basis to accomplish specific results—their goals

Organizations that make better, faster, and more effective decisions will have a decided competitive advantage.[8]

Decisions are where thinking and doing overlap. For this to happen within an organization, a decision must be logically consistent with the organization's risk management program and the willingness of the operations to agree they can do what is decided. Decision making involves a defined process and should be handled this way.

Decision making is connected with sense making and knowledge management (Figure 3.1).

It is focused on helping the organization meet its objectives in an uncertain world. Decision making uses knowledge to improve its ability to have successful outcomes. It also records what is learned during the decision-making process in the knowledge management system. Sense making takes the information from the scanning of the external operating environment and either stores it in the knowledge management system or tries to make sense of the information and make it available within the actual decision-making process. These two additional elements will be discussed in the next sections.

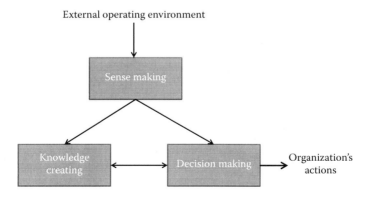

FIGURE 3.1
Relationships between decision making, sense making, and knowledge management. (Adapted from Choo, W.C., *The Knowing Organization: How Organizations Use Information to Construct Meaning, Create Knowledge and Make Decisions*, 2nd ed., Oxford University Press, New York, 2006.)

Decision-Making Techniques

There are a large number of different kinds of decision-making techniques. Organizations select from the techniques depending on styles that range from autocratic to unanimity-based decision making.[9] The criteria that shape decision making can be stated as follows[10]:

- A decision environment that may influence the decision style
- Complexity of the decision
- Value of the decision's desired outcome
- Alternative scenarios that have the potential to support the decision-making process
- Cognitive biases to the decision's selection and interpretation
- Quality requirements of the decision
- Personalities of those involved in decision making
- Time available to conduct the decision-making process
- Necessary level of commitment to or acceptance of the decision
- Impact on valued relationships that the choice of decision style may have

Several of these criteria may be in play at any given moment and amplify one another.

Decisions should be made by following a sequence of specific steps[11]:

- Organizations should know which decisions have a disproportionate impact on the organizational performance.
- Decision makers should determine when those decisions should happen.
- They should organize the structure of decision elements around sources of value for the organization.
- Decision makers should figure out what level of authority is needed, regardless of status, and elevate it to that level.
- They should align other parts of the organizational system, such as processes, data, and information, to support decision making and execution.
- Decision makers should help managers develop the skills and behaviors necessary to make decisions and translate them into action, promptly and diligently.

An organization's value is often seen as the sum of the decisions it makes and executes. Even good decisions can have unfavorable results. Most organizational decisions are not made in a logical, rational manner. Most decisions do not begin with careful analysis, followed by systematic analysis of alternatives and implementation of a solution. On the contrary, decision processes are usually characterized by conflict, coalition building, trial and error, speed, and mistakes. Leaders operate under many constraints that limit the rationality of their decisions. They use intuition as well as rational analysis in their decision making.[12]

Model for Making a Decision

This model demonstrates how risk can be managed as an integral part of managing an organization.[13]

Preparing to Make a Decision[14]

- Understand the purpose of what the organization is trying to achieve. What will the decision involve?
- Define the outcomes that would be desirable and their relation to helping the organization achieve its strategic objectives. Know how the outcomes will be measured.
- Consider who needs to be involved in the decision itself and the actions that will follow.
- Determine the criteria that will be used to make the decision. What is the basis for knowing whether the decision will be the correct one in light of the strategic objectives?
- Determine how to make the decision by defining the process and steps. What will be considered and what will be deemed irrelevant?

Making the Decision[15]

- Seek clarification and agreement of the key assumptions about the decision and the expected outcomes.
- Consider the range of things that might occur or are already present that would prevent the achievement of the intended outcomes.
- Do the same for the case of enhancing the achievement of the objectives.
- Consider the effect of those things that might occur or are already present on the achievement of the organization's strategic objectives.

- Reflect on the decision maker's recent performance under similar circumstances. Does this indicate the effectiveness of existing operating controls that are intended to enable the achievement of the organization's strategic objectives?
- What lessons can be learned by studying any recent activities that are similar to the focus of the current decision?

Acting after the Decision[16]

- Once the decision has been made, an action plan should be put into place to be able to measure the effect of the outcomes.
- Consider how to limit uncertainty associated with the desired outcomes by identifying new opportunities and threats and managing them.
- Consider the costs and benefits of alternatives to the path selected and decide which create the most value and provide the highest chance that the outcomes will be achieved.
- Consider the critical path to achieving the desired outcomes and allocate resources and accountabilities for all actions.
- Make certain that the decision-making and planning process is captured in the documented information and maintained within the knowledge management system.
- Establish and maintain a process for measuring performance and ensuring the completion of all the planned actions.

Monitoring and Reviewing Actions[17]

- Track progress and actions with appropriate oversight. Question any deviances from what was planned. Reallocate resources if required.
- Monitor the opportunities and threats (uncertainty) and scan the internal and external operating environment, if necessary, to see if there is additional uncertainty. Use sense making to determine the implications of any changes on what the organization is trying to achieve and its strategic objectives.
- Monitor the existing controls that are intended to enable the organization to achieve its strategic objectives.
- Conduct periodic reviews of those controls to propose appropriate opportunities for improvement.
- Monitor decision making and implementation as part of the organization member's or employee's performance evaluation while emphasizing and encouraging sound decision making.

- Verify continuous monitoring and periodic reviews with some level of internal audit.

Learning[18]

- Reach agreement on the extent to which the desired outcomes were achieved.
- Agree on whether the outcomes helped or hindered the ability of the organization to meet its strategic objectives.
- Agree on whether any consequential effects were successes or failures.
- Agree on the causes of the successes or failures and the roles of operational controls.
- Consider the implications for future decisions and define what the organization should learn.
- Capture all the information and learning in the knowledge management system.

Sense Making

Sense making is used to monitor changes in the external operating environment (context). It seeks to make sense of what is happening out there, why it is taking place, and what it means. Sense making is always perform retrospectively because it is not possible to make sense of events and actions until they have occurred and there has been sufficient time to construct their meaning with respect to decision making in the organization.

The process of sense making is usually put into play when there is some change occurring in the external operating environment. Through this process, the organization is attempting to understand the differences and determine the significance of those changes. There are three basic steps involved in sense making[19]:

- *Enactment*: The information is labeled, categorized, and connected together with information about stakeholders, events, and outcomes. Sometimes, the organization introduces new messages or actions into the external operating environment (e.g., distribute a document, hold a meeting, or create a website) and then refocuses its sense making on these activities. The enactment process segregates possible factors that the organization should clarify and take seriously.

- *Selection*: People look at the information they have gathered from the enactment and try to answer the question, what is going on here? This can be accomplished by overlaying new data with interpretations that have worked in explaining similar or related situations that have happened before. Interpretations are selected that provide the best fit with past understandings. A set of cause-and-effect explanations are created that could explain what was observed. These explanations have to be plausible, but they do not need to be the most accurate or the most complete.
- *Retention*: The products of successful sense making are retained in the knowledge management system for further use in the future.

Sense making is used to produce reasonable interpretations of data about change in the external operating environment of an organization. It can be used in conjunction with the scanning of the external operating environment. Use of the PESTLE tool, together with the enumeration of the opportunities and threats, produces information that is useful in the process of sense making. Engagement of the stakeholders found in the external operating environment offers opportunities for collaboration in the activity of both the scanning and the sense-making activity. In times of rapid change, it is imperative that organizations take advantage of these tools to help deal with the effects of uncertainty (opportunities and threats) on a frequent basis.

Knowledge Creation and Management

All organizations need to develop the capacity to continuously create new knowledge. Knowledge creation involves the management of tacit and explicit knowledge. It works best when there is a process that generates new knowledge by converting tacit knowledge into explicit knowledge. Tacit knowledge is the personal knowledge that members of the organization carry around in their minds. It is hard to formalize or communicate this knowledge to others. It consists of subjective "know-how," insights, and intuition that are derived from performing tasks over a long period of time. Explicit knowledge is formal knowledge that is easier to transmit between individuals and the organization as a whole. It may be codified in the form of work instructions, procedures, operational controls, rules, and other forms.[20]

The production of knowledge involves the conversion of tacit knowledge into explicit knowledge and back again. This typically takes place in four steps[21]:

1. *Socialization* is a process of acquiring tacit knowledge through sharing experiences. Tacit knowledge is transferred from an experienced person to another person by working side by side. This is built on the long-standing tradition of apprenticeships.

2. *Externalization* is a process of converting tacit knowledge into explicit concepts through the use of abstractions, metaphors, analogies, or models. This is what most people think of when talking about knowledge creation. It is most often used in the concept creation phase of new product development.

3. *Combination* is a process of creating explicit knowledge by finding and bringing together explicit knowledge from a number of sources. In this case, individuals exchange and combine their explicit knowledge through telephone conversations, meetings, emails, and other media forms. Existing information in computerized databases may be categorized, collated, and sorted to produce new explicit knowledge.

4. *Internalization* is a process of embodying explicit knowledge into tacit knowledge, internalizing the experiences gained through the other modes of knowledge creation into individuals' tacit knowledge assembled in the form of shared work practices.

These four modes of knowledge conversion follow each other in a continuous spiral of knowledge creation.

Transition to a Learning Organization

Outcomes of sense making are enacted environments or shared interpretations that construct the context of the organization. The knowledge creating model (Figure 3.1) depicts the organization as continuously tapping its knowledge to solve tough problems. Different forms of organizational knowledge are converted and combined in continuous cycles of innovation. The outcomes of the process include the development of new products and services, as well as new capabilities. The decision-making model sees the organization as a rational, objective-directed system. Decision makers search for alternatives, evaluate consequences, and commit to a course of action using the gathered information from sense making and knowledge creation. The outcome of decision making is the selection of courses of action that are intended to enable the organization to achieve its strategic objectives.[22]

Even while the rational decision-making framework is probably widely used over a wide range of different types of organizations, it is not uncommon for people to gather information for decisions and not use it. They ask for reports, but do not read them. Individuals fight for the right to take part in decisions, but then do not exercise that right. Policies are vigorously debated, but their implementation is met with indifference. Leaders spend little time in making decisions because they are engaged in meetings and conversations about the pending decisions.[23]

The knowing cycle model provides a structure and language that can be used to think about the use of information in organizations. The model does not specify a particular sequence or order (Figure 3.1). Instead, the three processes are simply interconnected, and there are many possible pathways along which the model can function. Sense making constructs the context, the frame of reference for knowledge creation and decision making. Knowledge creation expands organizational capabilities and introduces innovation. Decision making converts belief and capabilities into commitments to act.[24]

Learning

The organization should encourage improvement and innovation through learning. For the organization to attain sustained success, it is necessary to adopt "learning as an organization" and exploit learning that integrates the capabilities of individuals with those of the organization as a whole.[25]

Learning as an organization involves consideration of the following[26]:

- Collecting information from various internal and external events and sources, including success stories and lessons learned
- Gaining insights through in-depth analyses of the information that has been collected

Learning that integrates the capabilities of individuals with those of the organization is achieved by combining the knowledge, thinking patterns, and behavioral patterns of people with the values of the organization. This involves the consideration of[27]

- The organization's values based on its mission, vision, and strategies
- Supporting the development of learning models and demonstrating leadership through the behavior of leaders
- Stimulation of networking connectivity interactivity and sharing of knowledge both inside and outside of the organization
- Maintaining systems for learning and sharing of knowledge

- Recognizing, supporting, and rewarding the improvement of people's competence, through processes for learning and sharing of knowledge
- Appreciation of creativity supporting a diversity of the opinions of the different people in the organization

Rapid access to and use of such knowledge can enhance the organization's ability to manage and maintain its sustained success. For organizations wanting to remain relevant and thrive, learning better and faster is critically important. Organizational learning is neither possible nor sustainable without understanding what drives it.[28]

A learning organization values the role that learning can play in developing organizational effectiveness. It demonstrates this by having an inspiring vision for learning and a learning strategy that will support the organization in achieving its strategic objectives.[29]

A learning organization needs people who are intellectually curious about their work, who actively reflect on their experience, who develop experienced-based theories of change and continuously test these in practice with colleagues, and who use their understanding and initiative to contribute to knowledge development. Reflective practitioners understand their strengths and limitations and have a range of methods and approaches for knowledge management and learning, individually and collectively.[30]

Knowledge is a critical asset in every learning organization. Because learning is both a product of knowledge and its source, a learning organization recognizes that the two are inextricably linked and manages them accordingly.[31]

Learning organizations know how to harness the power of information and communication technologies without these technologies constraining knowledge management and learning. In a learning organization, information and communications technologies are used, among other purposes, to strengthen organizational identity; build and sustain learning communities; keep members, clients, customers, and others informed and aware of organizational developments; create unexpected, helpful connections between people and provide access to their knowledge and ideas; encourage innovation and creativity; share and learn from good practices and unintended outcomes; strengthen relationships; develop and access organizational memory; share methods and approaches; celebrate successes; identify internal sources of expertise; and connect with the outside world.[32]

For organizations that wish to embrace sustainability, it is important to build a learning organization. Irrespective of the approach taken to create a learning organization, isolated knowledge management initiatives will not last long. Only embedded, organization-wide activities to identify, create, store, share, and use knowledge can give knowledge management the role it requires to help with decision making and provide part of the foundation necessary for the sustainability journey.[33]

Despite the attention paid to strategic planning, the notion of strategic practice is relatively new. To create a strategy is relatively easy. However, to execute that strategy is quite difficult, since strategy needs to be synchronized with sense making, knowledge management, and decision making. Organizations should systematically review, evaluate, prioritize the sequence of, manage, redirect, and if necessary, even cancel strategic initiatives in a learning organization.[34]

Endnotes

1. ISO, 2009.
2. ISO, 2009.
3. ISO, 2013.
4. ISO, 2013.
5. ISO, 2013.
6. Choo, 2006.
7. Serrat, 2012a.
8. Serrat, 2012.
9. Serrat, 2012.
10. Serrat, 2012.
11. Serrat, 2012.
12. Daft, 2013.
13. Tophoff, 2015.
14. Tophoff, 2015.
15. Tophoff, 2015.
16. Tophoff, 2015.
17. Tophoff, 2015.
18. Tophoff, 2015.
19. Choo, 2006.
20. Choo, 2006.
21. Choo, 2006.
22. Choo, 2006.
23. Choo, 2006.
24. Choo, 2006.
25. ISO, 2009.
26. ISO, 2009.
27. ISO, 2009.
28. Serrat, 2009d.
29. Serrat, 2009d.
30. Serrat, 2009d.
31. Serrat, 2009d.
32. Serrat, 2009d.
33. Serrat, 2012a.
34. Serrat, 2009b.

4

Organizational System of Management

All organizations use some kind of system for managing their activities. Management systems may be used implicitly, or they may be used in an explicit manner. Formal management systems provide the organization with a set of processes and procedures that can be used to control its activities. A management system consists of a set of interrelated and interacting elements and a means for establishing a policy and operational objectives, as well as processes that can be used to achieve those objectives. From an organizational operations perspective, efficiency is gained by having an aligned system of management.[1] Any organization can choose to adopt a harmonized high-level structure[2] and use it to construct a "system of management" using elements of a large number of national and international standards.

At one level, leaders use a system of management as a process to help the organization meet its objectives. The system helps them provide a management service to the organization. At the same time, there are services and products that are using other processes for the operations of an organization. There are a large number of processes involved in a system of management and the operations. The process approach can be used to keep track of processes, activities, products, services, and the overall performance of the organization.

Creating an Aligned Management System

There are three categories of management systems found in organizations today:

1. Financial management and enterprise risk management (ERM)
2. Operating management system (OMS)
3. Functional management systems (e.g., quality, and environment, health, and safety)

These systems of management provide a level of management oversight for the organization's compliance with local, state, and federal laws and

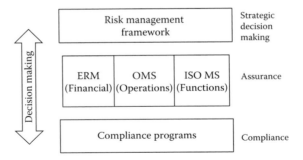

FIGURE 4.1
Role of management systems in an organization.

regulations and the work instructions for the organization's members or employees. The system of management is a useful means for helping management tend to the management of organizational risk[3] (Figure 4.1).

This approach may involve the following[4]:

- The application of the quality management system to control the activities involved in the provision of products and services to customers
- Dealing with operations that are regulated to prevent releases to the environment in the provision of products and services that may be of interest to external stakeholders
- Provision for the treatment of risk in activities that are managed by a health and safety standard
- Handling security risks associated with a cyber-attack on the organization's information systems
- Managing business continuity risks that provide a faster response to disruptive events
- Establishing controls to protect the organization's assets

The complexity of the activities common to a given organization and the uncertainty of its external operating environment create the need for a versatile system of management to enable the organization to meet its strategic objectives. There is an example of an aligned system of management in Appendix I. It lists many of the functional management systems currently in use. This chart describes the processes that can be taken from each and placed into the harmonized listing of processes, as described in the next section.

Processes in the System of Management

A system of management framework consists of the following documents[5]:

- *Standards*: Sections and descriptions, along with operational definitions
- *Policies*: Statement that provides the rules an organization uses to govern or constrain operations and its activities
- *Processes*: A high-level description of a large task or series of related tasks that describes "what happens" within the organization to create products and services that conform to the relevant standards in accordance with the policies of the organization
- *Procedures*: Detail "who" performs the work, "what" steps are performed, "when" the steps are performed, and "how" the procedure is performed

A standard is an agreed way of doing something. Standards represent a sense of "best practice," as they come from the wisdom of people with expertise in the subject matter who know the needs of the organizations they represent. They prepare standards working with others from all kinds of organizations, including regulators, associations, stakeholders, and customers. Standards are knowledge. They can help organizations meet their objectives (Figure 4.2).

A basic system of management is derived from the standards and has the following management processes associated with it:

- Scanning the external operating environment (context)
- Scanning the internal operating environment (context)
- Stakeholder engagement
- Determining the scope of the integrated system of management
- Creating the integrated system of management
- Characterizing the governance of the organization
- Creating the mandate and commitment from the leaders
- Establishing the policies of the component management systems
- Determining the roles, responsibilities, and authorities of the organization's members
- Using uncertainty analysis to address opportunities and threats
- Defining the organization's objectives and planning to achieve them
- Defining the resources needed to support the system of management

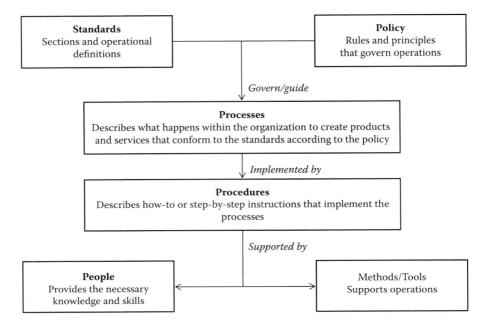

FIGURE 4.2
Documents of the system of management. (Adapted from ISO (International Organization for Standardization), Guidance on the concept and use of the process approach for management systems. ISO/TC 176/SC2/N 544R3, ISO, Geneva, 2008, http://www.ios.org/iso/04_concept_and-use-of-the-process-approach_for_management_systems.pdf, retrieved June 22, 2015.)

- Maintaining competence in all organization members or employees
- Creating an awareness program to keep all people current on the system of management
- Maintaining a system of internal and external communication
- Maintaining documented evidence necessary to support the system of management
- Developing and maintaining an operational planning and control system
- Developing and maintaining monitoring, measurement, analysis, and evaluation of the organization's ability to meet its objectives
- Providing for internal audits, self-evaluation, and maturity measurement
- Maintaining a regular management review process
- Maintaining a means for searching for nonconformities and taking corrective action when needed
- Maintaining a program of continual improvement

Using a process approach is a management strategy that helps give effect to the standard. Leaders seek to manage and control processes. The interaction between processes creates a coherent system of management.

People that work with systems of management will plan, implement, and maintain the systems with the processes listed above. It is important to find all the interactions between the processes in order to enhance the effectiveness of the processes. Each organization will implement the system differently to make sure that it is tailored to the organization. Just think of adding additional detail to the management processes so that each is a process that can be implemented by the organization's members or employees through the use of a procedure.

All the frameworks in use within an organization should be aligned to facilitate their interconnectivity potential while being available to improve decision making. Organizations should make sure that risk and uncertainty management is never a separate activity. It must be an integral component of the combination of all the management programs.

Organizations perform functions within each process to achieve their objectives. These functions are typically enacted through a set of tasks, each with their own objectives. Tasks are usually organized in the form of procedures. A procedure is a written sequence of instructions presented as a series of steps necessary to complete the task.[6]

It is important to first determine if the set of functions is "fit for purpose" in terms of achieving the objectives of the task. After determining the purpose of the task, task analysis is used to provide order to the procedure with a sufficient amount of information. Hierarchical task analysis starts with a high-level task and breaks it down into secondary-level steps, using lower-level steps to whatever degree of detail is appropriate and necessary.[7] In some cases, an operational breakdown matrix is used to provide additional information to the procedure. There are ways to test the procedures to make certain that they work properly by the people that are using them.

Managing for the Sustained Success of an Organization

In the international family of standards for quality professionals, a guide[8] was presented to assist organizations in achieving "sustained success," regardless of the size, type, and activity of the organizations. Sustained success was described as the ability of the organization to achieve and maintain its objectives over the long term.[9] Achieving sustained success should be possible if the organization follows the system of management for quality, as long as it is used based on the quality management principles[10]:

- Maintain a focus on the customer's requirements while striving to exceed expectations.

- Leaders need to establish the unity of purpose within the organization while creating and maintaining a work environment in which all people can become involved in helping to achieve the organization's objectives.
- People are the essence of an organization, and their engagement enables their abilities to be used for the organization's benefit.
- Desired results are achieved efficiently when activities and resources are managed using the process approach.
- Managing interrelated processes as a system contributes to the organization's effectiveness in achieving its strategic objectives.
- Continual improvement of the organization's overall performance is treated as an overarching objective.
- Effective decisions are based on facts obtained from the careful analysis of data and information.
- Organizations need to work with members of the value chains through mutually beneficial relationships that enhance the ability to create shared value.

It is easy to see how these principles contribute to sustained success, whether part of a quality program or used as an integral component of a risk management and sustainability effort.

An organization can achieve sustained success by paying close attention to its ever-changing external operating environment (see Chapter 8). Other elements of this approach include the following[11]:

- Creating a strategy and policy
- Managing resources
- Paying attention to people in the organization
- Working closely with suppliers and customers within the value chain
- Maintaining a suitable work environment
- Managing knowledge, information, and technology
- Managing processes
- Monitoring, measurement, analysis, and review
- Improvement, innovation, and learning

Each of these items is covered in Section II of this book (Figure 4.3).

Many organizations keep their systems of management separate from each other. However, the management of risk and uncertainty, along with sustainability, should be coordinated so that they can collectively contribute to the organization's core purpose and its strategic objectives. An organization

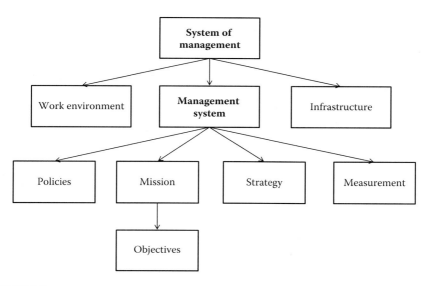

FIGURE 4.3
Relationship diagram of system of management. (Adapted from ISO (International Organization for Standardization), Quality management systems—Fundamentals and vocabulary, ISO/DIS 9000, ISO, Geneva, 2014.)

should enhance its resilience by ensuring that unnecessary overlap in processes and activities that wastes resources is avoided.[12]

Leaders should align operational activities through the achievement of coherence across the various systems of management. This can help an organization build the ability to adapt to changing conditions as they emerge. It can modify its structures, activities, and behaviors to adjust to new conditions while retaining its core purpose and objectives.

Process Approach for the Organization

The following methodology can be applied to manage the processes of the organization.[13] The approach starts with the identification of the processes of the organization:

- *Determine the purpose of the organization*: Create a mission statement.
- *Define the policies of the organization*: These will include risk management and sustainability.
- *Determine the sequence of the processes*: A sequence has been proposed above.

- *Define the process ownership*: Starts with the leaders and is cascaded down through the organization.
- *Define the process documentation*: This includes the process documentation and procedures.

Next, the organization needs to plan the process. It can do so with the following steps[14]:

- *Define the activities within the process*: These are necessary to achieve the intended outputs.
- *Define the monitoring and measurement requirements*: Provides the control and improvement of the processes and the intended process outputs.
- *Define the resources needed*: Determines the resources needed for the effective operation of the processes.
- *Verify the process against its planned objectives*: Confirms that the processes are consistent with the mission statement and the strategic objectives.

The third step involves the implementation and measurement of the process and its activities. The organization needs to provide the following assistance to the implementation effort[15]:

- Communication
- Awareness
- Training
- Management of change
- Direct involvement of the leaders
- Management review

The fourth step involves the analysis of the process. It is important to compare the results of process performance information with the requirements of the process as defined in the previous steps. This will help identify any need for corrective action. When corrective actions are needed, the method for implementing them must be clearly defined.

These four steps are often referred to as the plan–do–check–act (PDCA) methodology. The PDCA methodology is described in Chapter 12.

Embedding Sustainability

When people are talking about embedding sustainability into an organization, it should mean more than just engaging the employees. Sustainability must follow the example of risk management to be able to become an integral part of processes and be reliably cited in procedures. In Chapters 5 and 6, the example of risk management will be compared with sustainability implementation. The rest of the book will focus on helping organizations close the gap in how these important programs need to be a part of what every employee does every day.

Endnotes

1. ISO, 2013.
2. ISO, 2014.
3. ISO, 2013.
4. ISO, 2013.
5. Bandor, 2007.
6. Purdy, 2014.
7. Purdy, 2014.
8. ISO, 2009b.
9. ISO, 2009b.
10. ISO, 2009b.
11. ISO, 2009b.
12. BS, 2014.
13. ISO, 2008.
14. ISO, 2008.
15. ISO, 2008.

5

Risk Management

There are many definitions of risk. Most of the definitions focus on hazards, harm, and harmful events. Only one of these definitions is presented from the perspective of an organization. This definition covers not only something that might happen, but also how it will affect an organization's ability to meet its objectives in an uncertain world. No longer will risk be confined to harmful events. Instead, risk and its consequences could be positive (upside of risk) or negative (downside of risk). Risk is always focused on the organization's objectives. Opportunities and threats represent the effects of uncertainty.[1] This way of looking at risk from the perspective of an organization is now accepted by 50 countries representing more than 80% of the world's population.[2]

Understanding Organizational Risk

Risk is the effect of uncertainty on objectives.[3] This definition shifts the emphasis from an *event* (something happens) to the *effects* of uncertainty. Risk is associated with the strategic objectives. The effects of uncertainty can help the organization (opportunities) or hinder the organization (threats) from meeting the objectives. In the absence of uncertainty, risk comes from the execution of effective processes, efficient operations, and efficacious strategy.

After setting the strategic objectives (see Chapter 7), organizations have to address the internal and external factors that can generate uncertainty. This is accomplished when the organization prepares a scan of the external and internal operating environments. These operating environments are known as the context of the organization.

In the world of financial risk, organizations seek to avoid, control, or transfer risk to others (e.g., purchase insurance). Investors refer to the pursuit of opportunities as "risk and opportunity." The investors are pointing out that one has to take on more risk to realize the benefits of the opportunity. In the financial world, decision making causes risk in the same way that unfortunate events can cause risk. Poor decisions within the organization can prevent its ability to meet objectives. But this is not an event.

With the removal of the word *event* from the definition of risk, it is no longer correct to say that "risk has happened." When there has been an event, it is not proper to say the risk has "occurred." It is also not correct to describe a hazard or some other risk source as a risk or to characterize a risk as "positive" or "negative," although it would be valid to describe the consequences associated with a risk as beneficial or detrimental in terms of the organization's objectives.[4]

An *event* involves the occurrence or change of a particular set of consequences[5]:

- An event can be one or more occurrences that have several causes.
- An event can consist of something *not* happening.
- An event can sometimes be referred to as an incident or accident.
- An event without consequences can be referred to as a "near miss."

A *consequence* is the outcome of an event that affects the objectives[6]:

- An event can lead to a range of consequences.
- A consequence can be certain or uncertain and can have positive or negative effects on the objectives.
- Consequences can be expressed qualitatively or quantitatively.
- Initial consequences can escalate through *cascading effects*.

An *effect* is a deviation from the expected and can be positive or negative. Uncertainty has positive effects (opportunities) and negative effects (threats) with respect to an organization being able to meet its strategic objectives.

The *objectives* are the overarching outcomes that the organization is seeking to achieve. These effects are the highest expression of intent and purpose and typically reflect the organization's implicit goals, values, and imperatives.[7] Organizations establish responsible objectives; however, to achieve them, they must contend with the internal and external context of each operation and of all the other organizations found in the value chains. Objectives can have different aspects, such as economic, well-being, or environmental. In some organizations, the objectives mirror the three responsibilities of sustainability. We would expect this if the organization is seeking to balance its efforts to achieve sustainability; it must address the interests of the stakeholders, achieve its social license to operate, and address risk.

Understanding Uncertainty

Uncertainty originates in the internal and external context within which the organization operates. This can be uncertainty that[8]

- Is a consequence of underlying sociological, psychological, and cultural factors associated with human behavior

- Is produced by natural processes that are characterized by inherent variability (e.g., weather)
- Changes over time (e.g., due to competition, trends, new information, or changes in underlying factors)
- Is produced by the perception of uncertainty, which may vary between different parts of the organization and with its stakeholders

Uncertainty represents a deficiency of information that leads to an incomplete understanding of what can happen that would threaten the organization's ability to meet its objectives. Think of this as a recession, a severe storm, a devastating legal situation, or any number of things that could happen that would distract the organization from meeting its objectives. Uncertainty exists whenever the knowledge or understanding of an event, consequence, or likelihood is inadequate or incomplete.[9] Incomplete knowledge may involve information that, alone or in combination with other information,

- Is not available
- Is available, but is not accessible
- Is of unknown accuracy
- Is invalid or unreliable
- Involves factors whose relationship or interaction is not known

It may be possible to do something about some of these uncertainty elements, thus lowering the uncertainty.[10] The level of risk is expressed as the likelihood that particular consequences will be experienced. Consequences relate directly to strategic objectives. Consequences arise when something does or does not happen. Therefore, the likelihood being referred to here is not just that of the event occurring, but also the overall likelihood of experiencing the consequences that come from an event, and each will have its own likelihood.

When uncertainty is present, it creates *effects*. These effects can lead to a negative or positive deviation from the objectives that the organization seeks to achieve. Negative effects are often referred to as *threats*. Positive effects are referred to as *opportunities*. Risk consists of positive and negative effects.

An organization's objectives must be responsive to its internal and external stakeholders. The practice of sustainability seeks to have the organization establish responsible objectives to help it maintain its social license to operate. Sustainability has always been adept at seeking to find and create opportunities. Risk management is the larger influence and seeks to balance the threats and the opportunities. Sustainability is often faulted for being operated in a manner that is not embedded in the organization. The practice of sustainability can identify or create a lot of opportunities to help promote the organization's reputation. However, by working within the risk management program, sustainability can help the organization overcome the effects of uncertainty, thereby enhancing its chance of attaining its strategic objectives.

Risk Management Vocabulary

All activities of an organization involve risk. Organizations manage risk through a process of coordinated activities to direct and control opportunities and threats with regard to risk posed by each. It can also be seen as the driving force to achieve the organization's strategic objectives.

Risk management will always be enhanced by people understanding each other's perspectives. This means that everyone needs to understand how others view risk.

A common issue for both large and small organizations is a constraint on available resources for risk management and control activities. Therefore, it is important to keep these procedures as simple and straightforward as possible. The requirement of a risk management system is to have it be a fundamental part of how the organization operates every day, and not something that only specialists are allowed to opine on.[11]

Within a large parent organization, there are often multiple risk definitions in use, along with a number of different risk management programs. This presents a problem with regard to adopting a common risk language and no overarching program that will effectively manage risk.

Interaction between the often separate risk management fields within an organization (e.g., enterprise risk management, financial risk management, project risk management, safety and security management, business continuity management, and insurance management) can be ensured or improved, as the attention will not be primarily focused on setting and achieving the organization's objectives, taking risk into account.[12]

Stakeholder perception of risk can also vary to a great degree. This is caused by differences in assumptions, conceptions, and the needs, issues, or concerns as they relate to risk. It makes sense that the stakeholders would seek to make judgments of the acceptability of a risk based on their perception of risk. This is why it is important to engage the stakeholders in any risk management program. The organization needs to understand their interests and be sure that they are clear about the actual risks involved.

Adopting a Common Risk Language

A common risk language can be created by benchmarking the risk programs to the international definition of risk. This allows different disciplines, units, and geographies with distinct risk profiles to address the unique risks faced by including them in the context description. Risks common to all units in a hierarchical organization are handled in a strategic fashion. Risks unique to individual organizational units drive the unit-specific risk responses. It is important to realize that the international definition of risk is a high-level definition. It does not seek to preclude those definitions of the other units.

Some people advise against using the term *risk*. There are so many other terms that are used: *peril, loss, hazard, threat, harm, danger, difficulty, issue, obstacle, problem, luck, fortune, accident, possibility, chance, probability, likelihood, uncertainty, consequence, impact, outcome, level, event, occurrence, vulnerability, exposure, benefit, advantage, opportunity, windfall, prospect,* and so forth. All the more reason to have a common risk language! There are practitioners that still reject the international standard for risk. However, they often do not recognize that these other words also have a likelihood of misunderstanding when there is communication about risk.

It is important to avoid typologies of the areas of risk. It is more important to focus on understanding the fundamental processes driving uncertainty—hence risk—in organizations. Systems of management, organizational structure, people management, organizational culture, power, and conflict all have profound implications for organizational risk and uncertainty. The PESTLE analysis is helpful when conducted in the determination of the context. This helps an organization identify the uncertainty.

To be effective, risk needs to be regarded as important to the organization's strategic planning, management, and decision-making process. It is important to consider an organization's operating environment and determine how to integrate risk management with the governance arrangements.

The leading edge of risk management practice is addressing the management of uncertainty. This goes beyond perceived threats, opportunities, and their implications. It is about identifying and managing all the many sources of uncertainty that can give rise to and shape the perceptions of threats and opportunities. Uncertainty management implies exploring and understanding the origins of uncertainty before seeking to manage it. There can be no preconceptions about what is desirable or undesirable. Key concerns involve the understanding of where and why uncertainty is important in a given context, and where it is not important.

Risk Management Framework

A risk management framework is a means of managing uncertainty since the risk is focused on meeting the organization's strategic objectives (Figure 5.1). The framework does not describe a stand-alone set of activities, but rather what is happening within the routine work of the people in the organization. Everything in the framework is fully embedded into the way the organization operates every day. It is not a single document, nor is it a procedure, even though these elements can be important in the risk management framework. All the risk and uncertainty management activities in any organization can be compared with this framework.[13]

Risk management needs a strong and sustained commitment from the organization's leader to ensure it ongoing effectiveness. Leaders should do the following[14]:

- Create a risk and uncertainty management policy
- Ensure that the organization's culture is aligned with the policy
- Monitor and measure the risk management performance in a way that is similar to measuring other performance categories
- Embed risk and uncertainty management in the strategic objectives at all levels in the organization
- Align the goals of all members or employees with the risk and uncertainty management objectives
- Ensure legal and regulatory compliance

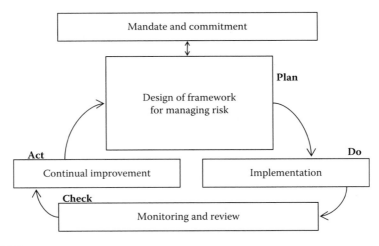

FIGURE 5.1
Design of the risk management framework. (From AS/NZS, Risk management guidelines, companion to AS/NZS ISO 31000:2009, HB 436, SAI Global Press, Sydney, 2013.)

- Ensure there are proper resources available for risk and uncertainty management
- Communicate the benefits of risk management to all stakeholders
- Ensure that the framework for managing risk and uncertainty remains appropriate

The design of the risk and uncertainty management framework must take into account the internal and external context of the organization. This is how it will identify the opportunities and threats (i.e., the effects of uncertainty). A number of general characteristics of the organization need to be considered in the design[15]:

- Organizational structure
- Governance practices
- Policies, internal standards, and operating model
- Contractual requirements
- Strategic and operational systems
- Capability and resources
- Knowledge, skills, and intellectual property
- Information systems and flows

These characteristics should be recorded so they can be referred to from time to time to detect any change that might require the framework to be adjusted.

Members or employees of the organization must have the appropriate competence for managing risk and uncertainty in order to be held accountable for their role in this risk management framework. The framework must be embedded in all the organization's practices and processes while keeping the contents associated with the framework relevant, effective, and efficient. Many organizations have a risk and uncertainty management plan to ensure that the design elements are embedded in all the organization's practices and processes. Often, this plan is part of the strategic plan.[16]

Once the risk and uncertainty framework has been designed, it is time to plan and execute the implementation of the elements so that the risk management process is routinely applied to decision making throughout the organization. When implementing risk and uncertainty management, the organization should do the following[17]:

- Define the strategy for implementing risk and uncertainty management
- Apply the risk management policy and process to the organization's processes

- Comply with legal and regulatory requirements
- Ensure the decision making is aligned with the outcomes of the risk and uncertainty management processes
- Provide awareness development activities
- Engage the stakeholders to ensure that the risk and uncertainty management framework remains appropriate and effective

Any weaknesses in the design or implementation of the risk management framework can lead to poor performance with respect to meeting the organization's strategic objectives. The framework needs to be monitored and reviewed to determine its effectiveness to support organizational performance in the following ways[18]:

- Measure risk management performance
- Periodically measure progress against the risk management plan
- Periodically review whether the design and its components are still appropriate
- Report on risk and uncertainty management progress and how well the risk management policy is being followed

Based on the results of monitoring and reviews, decisions should be made on how the risk and uncertainty management framework, policy, and plan can be improved. These decisions should lead to improvements in the organization's management of risk and its risk management culture.[19]

Embedding Risk Management

The focus of every organization should be on meeting its strategic objectives. This starts with the setting of responsible objectives that cascade from the top to the bottom of the organization (see Chapter 7). At this point, goals are set, along with action plans, to make sure the objectives are met. The management of risk should be embedded in the activities to manage the opportunities and threats associated with the effects of uncertainty found in the internal and external operating environment. Objectives need to be established and maintained mindful of these opportunities and threats. The people that are responsible for setting the objectives should also be responsible for managing the risk. However, the leaders of an organization know that everyone is responsible for the risk management within the structure of objectives and goal setting. Risk management should not be delegated to specialized risk practitioners in separate departments. Support staffs play a crucial role in assisting the "owners" with the effective management of risk.[20]

All the functional units in an organization (e.g., environmental, health, and safety; assets; quality; human resources; legal; sustainability; purchasing; and communications) must not have objectives that cannot be traced to the strategic objectives of the organization. Uncertainty should always be identified, assessed, responded to, reported, monitored, and reviewed in relation to the objectives the organization seeks to achieve, while giving consideration to changes occurring in the internal and external context.[21] When the members of an organization begin to understand the effects of uncertainty on the opportunities and threats, they can seek the guidance of the organization's governance to manage the risk associated with this uncertainty. Also, risk needs to be considered before a decision is made or actions have been initiated. This is true no matter where a decision is made within the organization.

Risk management must be tailored to the organization. Parent organizations must realize that there are differences in the internal and external contexts at each of their facilities or operations. The risk is similar to strategy. Both need to be adapted for ensuring success at the point where they are applied. Where risk management is embedded within the governance of the organization, the risk management function plays a number of important roles[22]:

- Facilitates proper risk management and internal control processes within all levels of the organization
- Serves as the custodian of the overall risk management and control frameworks
- Provides internal assurance on the effectiveness of risk management and internal control within the organization

The effects of uncertainty can never be completely eliminated. Organizations need to build both resilience and agility in all their activities that enable them to adequately respond to changes in circumstances or deal with the consequences of unforeseen events. All eyes need to be on the opportunities and threats and the uncertainty analysis that is used to prioritize them whenever there is a change in the internal or external context.

Risk management is built into the systems of management (see Chapter 12) used by organizations. Risk elements can be added or eliminated as needed. Once the management of risk is fully embedded as an integral part of the system of management, it becomes virtually invisible. Embedding of risk management into the organization's system of management helps the members or employees make intuitive decisions and take responsible and sustainable actions.

Endnotes

1. AS/NZS, 2013.
2. G31000, n.d.
3. ISO, 2009.
4. AS/NZS, 2013.
5. ISO, 2009.
6. ISO, 2009.
7. AS/NZS, 2013.
8. AS/NZS, 2013.
9. ISO, 2009.
10. AS/NZS, 2013.
11. SAI Global, 2005.
12. AS/NZS, 2013.
13. AS/NZS, 2013.
14. AS/NZS, 2013.
15. AS/NZS, 2013.
16. AS/NZS, 2013.
17. AS/NZS, 2013.
18. AS/NZS, 2013.
19. AS/NZS, 2013.
20. Tophoff, 2015.
21. Tophoff, 2015.
22. Tophoff, 2015.

6

Organizational Sustainability

Sustainability was first proposed as the goal of sustainable development. Sustainability is portrayed as having three dimensions—environmental, social, and economic. These dimensions are mutually reinforcing and interdependent. It is considered to be a misunderstanding to limit sustainability to a single dimension, such as climate change, nonsustainable resource depletion, or biodiversity. Sustainability is relevant to all levels of human activity, from the global level to the national, regional, community, organizational, and individual levels.[1] Since organizations are the basic building blocks of society, it is important to see how this perspective is different from the other definitions.

What Is Sustainability?

Sustainability refers to a state of the global system with a focus on the environmental, social, and economic subsystems, in which the needs of the present are met without compromising the ability of future generations to meet their needs. From this perspective, sustainability is a characteristic of the planet as a whole and not of any particular activity or organization.

Sustainable development addresses the activities and products of particular organizations (or nations, regions, and communities) and the ability to engage in such development in a manner that contributes to sustainability. Development is needed to meet the needs of both present and future generations, and is essential to sustainability.[2]

Social responsibility encompasses an organization's responsibility for the impact of its decisions and activities on the environment, society, and the economy. As such, social responsibility is the organization's contribution to sustainable development and sustainability. Social responsibility is applicable to all organizations (not just corporations) as they recognize that they also have a responsibility to contribute to current sustainable development and future sustainability.[3]

Sustainability is often described as the goal of sustainable development. Understanding and achieving a balance between environmental, social, and economic systems (ideally in mutually supporting ways) is considered essential for making progress toward achieving sustainability. Many experts believe that the achievement of sustainability needs to be recognized as one of the most important considerations in all human activities. To summarize, the term *sustainable development* is used to describe development that leads to sustainability. The term *social responsibility* is used to describe how an organization can contribute to sustainable development.[4] It would make sense that as the organization begins to contribute significantly to sustainable development and sustainability, it may be referred to as a sustainable organization. This distinction can remove some of the confusion over using the term *social responsibility* for the organization.

Organizational Sustainability

From the perspective of an organization, sustainability is the capability of an organization to transparently manage its responsibilities for environmental stewardship, social well-being, and economic shared value over the long term while being held accountable by its stakeholders.[5] This definition is actionable with any organization at the community level. The definition can form the bottom-up feedback loop for any definition of corporate sustainability or corporate social responsibility. No euphemisms, no slogans, no colors, and no initiatives!

Each of the responsibilities can be defined by best practices associated with social responsibility.[6]

Some organizations define responsibilities in general terms[7]:

- Doing what you have committed to do
- Always giving your best effort
- Being held accountable for your choices
- Helping others when they need help
- Being fair
- Helping to make a better world

To reinforce these responsibilities and make them more specific to daily activities, an organization may create a "code of conduct" that outlines expectations for how responsibility will be embedded into what employees do every day. Organizations may also specify responsibility as part of their core values to ensure that "acting responsibly" is part of the organization's culture.[8] These codes of conduct typically state that the organization should[9]

- Be accountable for its impacts on the environment, society, and the economy
- Be transparent in its decisions and activities that impact its responsibilities

- Behave ethically
- Respect, consider, and respond to the interests of its stakeholders
- Accept that respect for the rule of law is mandatory

Each organization will have its own list of significant sustainability responsibilities. While they are often divided up by categories, please remember that each of the responsibilities is interconnected with the other two.

To be effective, the code of conduct should consider a number of items[10]:

- Maintaining evidence of compliance with relevant local, regional, and federal laws
- Understanding the consequences of noncompliance with the laws and regulations
- Effectively managing the elements of the code of conduct
- Maintaining the integrity and reputation of the organization
- Adhering to aspirational values in line with sustainability
- Dealing with conflicts of interest and confidentiality
- Being responsible for engaging with external stakeholders
- Maintaining nondiscriminatory practices
- Paying attention to how members or employees are treated
- Conditions of membership or employment
- Using and accounting for organizational resources
- Conditions within the organization are safe and hygienic
- Paying attention to occupational health and safety
- Acting as a steward of the environment and a good citizen of the community

An organization's commitment to the code of conduct should include the benefits and importance of having such a code or honoring a supplier code of conduct. The code needs to be an integral component of the framework developed to manage risk and sustainability within the organization.

For an organization to practice *environmental stewardship*, its activities, processes, decisions, products, and services should strive to have no net negative consequence to the environment. Environmental responsibility involves the following[11,12]:

- Having effective processes for all operations associated with products or services and a system of management to provide management oversight of the processes and operations (e.g., environmental, assets, energy management, business continuity, and sustainable development)

- Enhancing the productivity of natural resources—use only what is needed, use it efficiently with a focus on reducing or eliminating waste, and be aware of issues with the management of waste that is not eliminated, including the reuse of products at their end of life
- Being mindful of energy use and climate change and not simply switching to renewable energy, which has significant Scope 3 greenhouse gas emissions or large amounts of embodied energy in the fabrication, installation, operation, and maintenance of the technology
- Addressing the stewardship of the natural habitat and its biodiversity in the neighborhood, the community, and in areas affected by the value chain partners
- Paying attention to widespread contamination of the environment and in human tissue of large numbers of biopersistent chemicals
- Considering the interests of the organization's external stakeholders identified in its external operating environment

In the case of *social well-being*, the organization should seek to avoid negative consequences to society as a whole, with particular emphasis on its members, employees, and other people directly impacted by its activities, processes, decisions, products, and services. All its suppliers should be contractually obligated to follow its supplier code of conduct or have systems of management in place to facilitate the progress to sustainable development. Social well-being includes the following[13,14]:

- Respect for human rights by having the organization exercise due diligence in seeking to determine where human rights issues may arise within its value chain
- Responsibility for its labor practices, both in its own operations and where it has a sphere of influence in the value chain
- Having a system of management (e.g., health and safety; governance, risk, and compliance; social responsibility; and risk management) in place to facilitate the attainment of social well-being
- Adoption of fair operating practices to deal ethically with other organizations, including the prevention of corruption, responsible participation in the political process, respect of property rights, and promoting responsibility in its sphere of influence
- Involvement in community development and participating in local education and culture, public health, literacy, social investment, and quality of life

Each organization should use its control or sphere of influence in partnership with other local organizations and its value chain partners to address *economic issues* and their interrelationships with the other two sustainability responsibilities[15]:

- Employment in the community
- Poverty and similar needs
- Local business climate
- Income levels
- Economic performance and community development
- Use of technology and innovation
- Value and supply chain prosperity
- Maintenance of the social license to operate
- Working with other local organizations to promote the value of organizational sustainability

Here are items that could be included in an organization's list of responsibilities[16]:

- Continually improve the resource productivity of the operations
- Eliminate wastes of all kinds
- Pay attention to the prevention side of the activity rather than using recycling or controls
- Manage energy to respect the need for climate change mitigation
- Protect natural habitats and biodiversity
- Consider environmental impacts in areas under the organization's control and within its sphere of influence
- Protect human rights with an evaluation of the entire value chain
- Ensure fair operating practices
- Assess labor practices, including health and safety
- Evaluate consumer issues associated with products and services
- Optimize community involvement
- Consider social impacts and license to operate in areas under the organization's control and within its sphere of influence
- Contribute to the community's development
- Look for opportunities to share value within the community
- Consider community shared value impacts in areas under the organization's control and within its sphere of influence

Attention to these items should help an organization operate responsibly.

Operating Responsibly

The concept of operating responsibly is at the core of an organization's sustainability program. As in the case of risk management, the three responsibilities must be integrated with each other and embedded in the organization's activities, processes, decisions, services, and products. Responsibility is seen as a balanced approach for organizations to address environmental, social, and economic issues in a way that aims to benefit people, communities, and society.[17] Organizations are responsible for the consequences of their activities and decisions through their transparent and ethical behaviors. The responsibility extends to the customers, neighborhood, community, society, and environment. Exercising an organization's responsibility involves many aspects of its operations[18]:

- Contribution to sustainable development, including health and welfare of the community and society
- Active engagement with stakeholders to determine their interests in the organization and its products and services
- Operating in a manner that complies with applicable laws and is consistent with the international norms of behavior
- Integrating sustainability throughout the organization and practicing it in relationships that are within the organization's control or sphere of influence

A number of specific relationships guide the responsibilities of an organization. They include relationships between the organization and society, between the organization and it stakeholders, and between the stakeholders and society. All these relationships affect the operation of organizations at the community level.[19]

First, and most importantly, all organizations should have a relationship with the community. Every organization needs to understand how its activities, processes, decisions, products, and services affect the community. Organizations often support the community as a great place for their employees or members to live. Often, the suppliers have operations in the community as well.

Second, organizations have a relationship with their stakeholders—both inside and external to the organization. Just as each organization engages with the internal stakeholders, it is important to extend the engagement beyond the customer to other external stakeholders. This dialogue with stakeholders should be face-to-face and interactive, and occur over long periods of time. The interests of the stakeholders need to be understood and acknowledged. If there are issues with stakeholders, some form of mediation should be considered. This will enable engagement to dominate the agenda with all stakeholders.

Third, the organization's stakeholders have a relationship with the larger community (state, province, regional authority, and federal authority). Since stakeholders can be associated with diverse groups, it is possible that some of their interests are not consistent with the expectations of the community at large. Stakeholder interests must be carefully balanced across a broad spectrum of interests. It is important that the stakeholder organizations realize that the interests of other stakeholders may conflict with their own.

Organizations must understand how these relationships can complicate their ability to maintain their social license to operate (see Chapter 9). To some degree, local organizations have always conducted their activities with particular awareness of their relationship to the community. However, with the range of communication methods available today, it is even more important that they pay particular attention to these relationships. The organization must responsibly decide how it will embed sustainability into its operations, rather than focusing on "initiatives" that compete with its core day-to-day operational activity.

Embedding Sustainability

Just as with risk management, organizational sustainability should be part of what every member or employee does every day. At the parent organization level, claims are often made that sustainability is embedded within the entire organization's structure and functions. The reality is that very few parent organizations have fully embedded or integrated sustainability into the way they operate day in and day out. Sustainability is frequently operated as a separate program with its own goals that are not aligned with the organization's strategic objectives. Many of these goals are designed to appease outside interests and not for the point of operating in a stewardship mode seeking to prevent creating environmental, social, and economic problems.

There are two different forms of embedding sustainability. The first involves making sustainability and the responsibilities associated with sustainability part of the work instructions and operational controls of everyone in the organization. Sustainability would be clearly part of what they do and not practiced solely as a separate activity (e.g., green team initiatives). A second way of embedding sustainability is to make its considerations part of every decision made at all levels of the organization. In either case, there needs to be a close connection between sustainability and the strategic objectives of the organization. It is also important for there to be shared value between the stakeholders and the organizations. This is more complicated at the parent organization scale than it is at the organizational level. The strategy comes from the mission statement in terms of the strategic objectives. These objectives are cascaded down to the lower levels of the organization. Workers have goals and an action plan to achieve the goals using the guidance and structure of the strategic objectives covering their work. The realization of the goals at each level in the organization can be compared with the objectives to see if there is value created over and above the meeting of the objectives. Sustainability would need to be included within the strategic objectives, as well as the focus of the action plans associated with every worker's goals.

It is important that the members or workers in every organization understand their organization's strategic objectives. These objectives need to be transparent to both internal and external stakeholders. When goals are established, many of the responsibilities listed above can be incorporated as potential means for creating value and ensuring the effectiveness of the stewardship approach. The effects of uncertainty must be dealt with if the organization wants to meet its strategic objectives.

Managing Opportunities and Threats

Every organization operates in an uncertain world. This presents the operations with opportunities and threats (i.e., the effects of uncertainty). The external influences and factors are responsible for creating the opportunities and treats (see Chapter 8). As part of the goal setting and action planning, any significant opportunity or threat needs to be examined and managed through the action planning process. By controlling the effects of uncertainty, the organization is more likely to be on the upside of risk. This creates value for the organization. If there are unattended threats or unrealized opportunities, the organization is more likely to be on the downside of the risk. Risk management and uncertainty analysis must be a part of the sustainability program once it is integrated into the operation of the organization.

Organizational sustainability has a lot of moving parts, but they are much easier to control at the organizational level. The rest of this book will show how this is accomplished.

Endnotes

1. ISO, 2014d.
2. ISO, 2014d.

3. ISO, 2014d.
4. ISO, 2014d.
5. Pojasek, 2012.
6. Pojasek, 2010.
7. Pojasek, 2010.
8. Pojasek, 2010.
9. ISO, 2010.
10. AS, 2003a.
11. ISO, 2014d.
12. Pojasek, 2012.
13. Pojasek, 2012.
14. ISO, 2014.
15. ISO, 2014.
16. Pojasek, 2012.
17. ISO, 2010.
18. ISO, 2010.
19. ISO, 2010.

Section II

Structure for Planning and Implementing Organizational Sustainability

Most successful organizations use a plan–do–check–act (PDCA) sequence to implement and maintain programs to improve quality; environmental protection, health, and safety; and other functional disciplines. Some PDCA programs are implicitly stated, and some are explicitly defined in a system of management. Chapters 7 through 14 provide information and guidance to help any organization use the "plan" and "do" (i.e., implementation) as a means of implementing and embedding sustainability into an organization. The information is based on international best practices that are included in the endnotes section for each chapter and linked to the references at the end of the book.

The reader will learn what is important to have in the sustainability program and can decide *how* best to make that happen in a particular organization. The popular slogan for this activity is "Say what you do! Do what you say! Do it effectively! Be able to prove it!" However, it takes some practice to make this happen.

In Appendix II, there is a case for implementing a sustainability program at a virtual hotel. It is not a bad start, but it does not use all of the materials presented in these chapters. The case also needs to be examined from the point of view of whether the essential questions (presented at the end of each chapter) have been addressed in the case. Since this is a practical step-by-step guide, this information should help the sustainability practitioner implement sustainability into any organization and see that it is embedded within what every person in that organization does every day. Appendix III offers a means to practice the "check" and "act" elements of PDCA.

7

An Organization's Objectives and Goals

To achieve sustainable success, an organization must meet its overarching objectives over the long term while operating in an uncertain world. When developing a sustainability program, it is important to understand the organization's objectives and how they were established. As with other terms, there is disagreement regarding the difference between "objectives" and "goals." Many people continue to use these terms interchangeably. There is also disagreement regarding the purpose and content of a mission statement—the principal source of the organization's objectives. The focus of this chapter is on setting objectives and goals in an organization. The literature surveyed in the previous six foundation chapters will be used to maintain the perspective of the organization.

Organization's Mission Statement

The mission statement is widely regarded as an explicit statement of the reason for the organization's existence and what it is meant to accomplish. Mission statements are typically focused on a 5- to 10-year time frame and should

- Separate what is important to the organization's sustainable success from what is not as important
- Clearly state the customers, clients, or other persons and organizations that are served and how they are served
- Communicate the organization's "looking forward" position to its stakeholders

For a private business, the mission is focused on its products or services. In the public sector, the mission statement focuses on what the organization is trying to accomplish. Most not-for-profit organizations are established to start something new or stop something that they find objectionable, such as protecting a wetland area or eliminating the emission of greenhouse gases. In all cases, the mission statement provides a foundation for the strategic planning process and the management of the processes and activities within the organization. A mission statement is very important to an organization because it helps management increase the probability that the organization will achieve sustainable success over time.

A mission statement is different from a vision statement. Most regard the mission statement to be the *cause* and the vision statement to be the *effect*. Looking at this relationship in a different way, the mission statement is something to be accomplished and the vision statement something to be pursued for that accomplishment.

Most small organizations do not have a written mission statement, but rely on the implicit understanding of what is most important to the organization—what it stands for. Preparing a written mission statement enables the organization to explicitly state its purpose for anyone with an interest. The written mission statement should provide the reader with a brief summary of the organization's principles and culture[1] (see Chapter 2). There will be a keen interest by internal and external stakeholders in knowing how the organization's values will determine how the mission is executed moving forward. Sustainability may be embedded into the operational objectives, along with the organization's values, to become part of how the organization operates every day.

Many organizations seek to be a good neighbor and a positive contributor to the community. Their work should not cause harm in the neighborhood. Organizations need members or employees to create the products and services associated with their operations and mission. If the community suffers from environmental, social, and economic problems, the organization is likely to have trouble recruiting members and retaining them at this location. It is always easier to attract and retain members in an organization when a community is growing in a sustainable manner. Having a beneficial relationship with the local community is vital to securing and maintaining its "social license to operate." By enabling beneficial collaborations and partnerships with the community and other organizations in that community, it is possible to aspire to values that will set the organization apart from its competitors. Sustainability and social responsibility provide a means for the organization to build long-term relationships with local suppliers and customers while enhancing its reputation as a good neighbor in the community.

An organization can use its mission statement to communicate to its internal and external stakeholders a good sense of what it is seeking to achieve in the area of sustainability. In this manner, the organization can communicate its legitimacy to the community and seek new members, employees, or trusted partners that can identify with its stated purpose. Some organizations focus their mission statement on making a profit. This kind of statement may not be viewed favorably by some external stakeholders because it will seem that the organization places profits above addressing the interests of the stakeholders. A sustainability program helps create responsible mission statements and objectives that demonstrate the value of its environmental stewardship, its focus on social well-being, and the shared value associated with these contributions. However, these contributions must not be seen as something that is provided separately by an organization that struggles with coming to terms with its place and role within the community.

Organization's Objectives

Once the mission statement is explicitly stated, it is important to link the mission to the objectives[2] of the organization. This will help provide the means for accomplishing the mission. At the highest organizational level, strategic objectives are statements of broad intent that are strategically linked to the mission statement. Like the mission statement, these strategic objectives typically have a window of accomplishment of approximately 5–10 years. Ideally, an organization should have between three and five strategic objectives derived from its mission statement. Each strategic objective should be concise, specific, and able to be understood by the stakeholders. It is considered to be good practice to state each strategic objective in a single sentence with fewer than 25 words. Many involved in the field of organizational development consider objectives to be "continuous," while the goals have a clear beginning and end associated with them.[3] In this way, the objectives provide the bridge between an inspirational mission and the clearly stated goals that will provide a feedback loop from the bottom of the organization up to provide some proof that the strategic objectives are on target.

Objective setting does not end at the strategic level. Objectives must also be prepared for each of the organization's operating levels, local operations, and individual work units. At the operating level, objectives establish the "outcomes" sought through the processes and operations and explain what the organization is seeking to accomplish at these operating levels. Objective setting is a top-down process that must extend from the strategy-setting level to the lowest operating levels of the organization (Figure 7.1). The objectives at the lowest operating levels must be linked to objectives in the next higher level and be directly traceable to the strategic objectives. Objectives at the lowest levels of operations help to motivate the members or employees to work effectively to help the organization realize its overarching strategic objectives. The operational objectives are specific and short term, so they can be used in the operation's day-to-day activities. The relationship between an organization's overarching strategic objectives and its operational objectives is a key factor in its ability to meet its objectives in an uncertain world. Long-term organizational success is only likely to happen if short-term operational activities are consistent with the long-term strategic intentions. This is important to keep in mind when embedding the sustainability program into this objective structure. Sustainability objectives must be aligned with the operating objectives and directly support the strategic objectives.

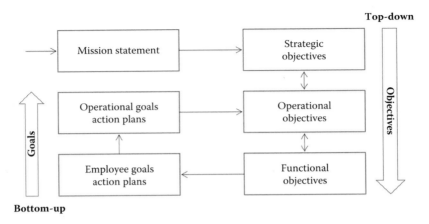

FIGURE 7.1
Relationship between objectives and goals in an organization.

Goal Setting and Operational Execution

An organization's strategy is implemented at the operating level by defining work goals and developing action plans that help people reach these goals. Goals provide a bottom-up feedback loop to test the top-down objectives (Figure 7.1). They state the specific measurable results that specify how much or what will be accomplished by when. Goals linked to action plans let management know whether operations are meeting their immediate operational objectives. Each work goal is supported by an action plan.[4] These action plans provide a prioritized set of activities that must be achieved in order for the work goal to be realized. Every action is subject to the question, how does this effort help the organization's operational and strategic objectives? This question helps the member or employee think of how the organization operates with its processes and activities. No activity performed within the organization should be a stand-alone effort that is separate from the main activities and processes of the organization.[5]

Every member of an organization, whether in a supporting position or an operational role at any level, should have one or two goals to work toward. There is evidence that suggests the chance of effectively achieving two or three goals is high; however, the more goals that someone tries to achieve at one time, the more likely it will be that none of the goals are attained.[6] The suggestion is to narrow the focus of each organization member to one or two "wildly important" activity goals in the short term and consistently invest the person's time and effort in achieving those goals. These wildly important goals will focus on the significant opportunities and threats that are prioritized by using risk assessment (see Chapter 11). It is not about selecting goals that provide sustainability results or meet other needs. Rather, it is about setting goals so fundamental to the mission of the organization that achieving them ensures that the strategic objectives will be realized. Once the activity goals are selected, a number of lead and lag measures are used to track the progress following the action plan. The measures are carefully tracked to help the members or employees create a "cadence of accountability."[7]

Sustainability programs are often operated as a series of initiatives with their own goals, which can lead to a common and ineffective pattern of organizations setting objectives and goals as separate efforts or interchanging the two terms so there is no organized functioning of the organizational structure that is presented in this chapter. When considering that the organization is operating in an uncertain world, it is prudent to coordinate the use of objectives and goals with the risk management efforts (see Chapter 5). The strategy should always be in line with the mission statement and the stakeholder interests. Organizations must be capable of creating an efficacious strategy that is able to effectively address the mission statement through the auspices of its overarching strategic objectives.[8]

Writing or Revising an Organization's Mission Statement

Many larger organizations have written a mission statement. In many cases, the mission statement comes from a parent organization at the top of a hierarchal structure. If there is an existing mission statement, it can be evaluated or revised prior to the process of creating a set of strategic objectives. The mission statement process starts with a review or the drafting of the mission statement to make sure that it addresses four sets of questions[9]:

- Who are we? What do we do? What are our products and services?
- Who do we serve? Who finds these products and services of value?
- Why are we here? What value do we provide? What business problem, human need, or desire do our products and services fulfill? Have we identified the most important values?
- What principles or beliefs guide the organization?

There are many guides to writing mission statements on the Internet. Remember that a well-crafted mission statement[10]

- Expresses the organization's purpose in a manner that inspires support and ongoing commitment to provide the organization with a unique identity
- Is articulated in a way that is convincing and easy to understand

- Uses proactive verbs to describe what the organization does
- Is free of unnecessary jargon
- Is short enough to enable stakeholders to easily recognize it

A well-written mission statement is an important prerequisite for the development of the strategic objectives.

Developing the Objectives

To develop the strategic, high-level objectives, the drafting team needs to ask the following questions:

- What are the three to six areas in which the organization will continue to be actively involved over the next 5–10 years?
- What areas need attention to accomplish what is found in the mission statement?
- Do the objectives convert the mission statement into action?
- Do the objectives help to sustain the organization's competitive advantage?
- Is there a risk management program in place for prioritizing the opportunities and threats found in the organization's internal and external operating environment (context)?

The objectives of a parent organization affect the objectives throughout an organization. Business or divisional objectives are derived from the corporate objectives, but are specific to these levels of the organization. Department objectives are derived from the divisional objectives so that the linkage back to the strategic objectives can be maintained. This progression is repeated down to the individual member or employee level. It does not matter whether the organization is a business or a not-for-profit. Similarly, the time frame for objectives can range from 5–10 years for long-term strategic objectives to 1 year for an objective at an individual level.[11] Every organization needs to have more than one objective at the strategic level. Objectives are always derived from the mission statement. A sustainability program can set strategic objectives only if there is a focus on sustainability that is embedded within the operational or functional levels, which support the strategic objectives at the highest level in the organization.

The mission and the strategic objectives should not be considered to be unchangeable.[12] Mission statements form the foundation for objectives, and both can change over time. There are factors in the external operating environment that can create opportunities and threats as the uncertainty of the operating environment is changed. These effects will be discussed in Chapter 8.

Developing Goals and Action Plans

Ideally, both supervisors and the members or employees at the lowest operational levels should participate in defining the goals. The supervisor helps provide clarity about the operational objectives and their role in setting the strategic direction. The employees should work as a team to develop the goals as an expression of their commitment to the strategic direction. Achieving all the individual goals in an area should make it possible to achieve the operational objective that has been assigned to that supervisor. Each member or employee would have one or two goals that are clearly stated and measurable.

There are a number of important questions that must be addressed within this goal-setting process[13]:

- What element of the performance of our work can best help achieve the operational goals?
- What are our greatest strengths that can be leveraged to ensure that the goal is achieved?
- What area of our past performance needs to be improved to ensure the goal is achieved?

The focus of these questions should be on defining the goals. It is critical to identify the goals that promise the greatest positive consequence for the operational objective that supports the strategic objectives. The answers to these questions should therefore be ranked by consequence—for the organization as a whole, not for the area's performance (see Chapter 11).

Once there are a few candidates, the team tests the goal against the so-called SMART elements[14]:

- Specific
- Measurable
- Achievable
- Realistic
- Timely

Once these elements are satisfied, the worker or team can assist the supervisor in defining the goal so that an action plan can be prepared.

An action plan provides the means to make sure that the achievement of the goal will help the organization meet its objective. The action plan needs to be complete, clear, and current. It should include information and ideas that were gathered during the planning of the goals. The following information is needed[15]:

- Describe the goals to be achieved.
- Describe the specific, measurable, and attainable outcome-based goals.
- What tasks, actions, or activities will occur to accomplish the goal?
- Who will conduct and be held responsible for these actions?
- When will the activity take place? Provide an appropriate timeline.
- Secure the allocation of the resources (e.g., money and people) needed to conduct the activity.
- How will the activity, along with its successes and opportunities for improvement, be communicated within the organization?

If the organization does not have a formal action planning process, there are some good examples to follow.[16,17] Management should review the action plans on a regular basis as part of their effort to ensure that the organization achieves its objectives.

Sustainability Goals and Objectives

If the sustainability program is embedded within the organization, the operational objectives will require sustainability goals to be established as described above. Alternatively, the sustainability goals can be accommodated as part of one of the organization's other structural elements, as described in the following chapters. Means for monitoring and measuring these activities can be found in Section III of this book.

Uncertainty Affects the Objectives and Goals

The "top-down" objectives and "bottom-up" goals guide the system of processes and activities to help the organization meet its objectives for

> ## BOX 7.1   ESSENTIAL QUESTIONS FOR SETTING SUSTAINABILITY GOALS
>
> Why are sustainability goals established without regard for the organization's strategic objectives and goals?
>
> How can the sustainability goals be used to influence the organization's strategic objectives? How can the sustainability initiatives be defined around the employees' goals?
>
> Why are sustainability goals focused on producing results rather than on helping the organization meet its strategic objectives?
>
> How can the stakeholders be affected when the organization meets its strategic objectives?
>
> What happens when the results of the employees meeting their goals do not contribute to meeting the operational objectives at that level?

sustainable success. An organization's operations are the day-to-day activities that are responsible for delivering the strategy. There are events and decisions that happen along the way that create opportunities and threats for the operations. These opportunities and threats are identified by characterizing the organization's context at the operating level.

Essential questions for addressing sustainability within the organizational objectives and goals can be found in Box 7.1.

Endnotes

1. Davis, n.d.
2. ISO, 2014.
3. Davis, n.d.
4. Pojasek, 1997.
5. ISO, 2009.
6. McChesney et al., 2012.
7. McChesney et al., 2012.
8. Hopkin, 2012.
9. NORC, n.d.
10. NORC, n.d.
11. Barnat, 2014.
12. Barnat, 2014.

13. McChesney et al., 2012.
14. Davis, n.d.
15. Community Tool Box, n.d.
16. Community Tool Box, n.d.
17. Pojasek, 1997.

8

Organization's Internal and External Context

Management of an organization involves a large number of coordinated activities to control its activities, processes, operations, products, and services as it pursues its strategic objectives.

Risk management is a component of organizational management involving coordinated activities associated with the effect of uncertainty on those objectives. Effective risk management is fundamental to the success of the organization, as it focuses on its performance against the objectives.[1] An organization's performance in relation to society and the economy, as well as its impact on the environment, has become a critical part of measuring its overall performance and its ability to continue operating effectively. Because risk is the effect of uncertainty on achieving strategic objectives, it is important to examine the influences and factors in the internal and external context that can have an effect on the organization's objectives.

Context of the Organization

Organizations operate in an uncertain world. Whenever an organization seeks to meet its objectives, there is a chance that everything will not go according to plan. It is possible that the organization will not achieve its objectives even if the objectives were carefully planned. What lies between the organization and its objectives is uncertainty. Uncertainty represents a deficiency of information that leads to an incomplete understanding of what can happen that would either threaten or enhance the organization's ability to meet its objectives.

The operating environment within which the organization operates includes many sources of uncertainty. This uncertainty creates "effects." These effects can be negative or positive. Negative effects are referred to as "threats." Positive effects are referred to as "opportunities." These effects create positive or negative consequences[2] that affect the organization's ability to meet its strategic objectives. Consequences are the outcomes of an event or decision affecting the organization's objectives. Effective risk management helps the organization understand its opportunities and threats, and address them as appropriate to manage the consequences and thereby maximize its chance of achieving its objectives by managing the uncertainty.

It is important to remember that risk is associated with the performance of the strategic objectives of the organization.

Internal Context

An organization's internal context is the operational environment within which it seeks to achieve its objectives. The internal context includes anything that the organization has control over or where it has a sphere of influence. The sphere of influence is an important concept. It represents the range and extent of political, contractual, economic, or other relationships through which an organization has the ability to affect the decisions or activities of individuals or organizations.[3] The internal context can include the following[4]:

- Governance, organizational structure (including relationships with a parent organization), roles, and accountabilities
- Policies, objectives, and strategy
- Operational capabilities understood in terms of processes, operations, and resources
- Decision making, knowledge, and sense making (see Chapter 3)
- Engagement with the internal stakeholders
- Relationships with other organizations within its value chain
- Organization's culture (see Chapter 2)
- Standards, guidelines, and operating models adopted by the organization
- Honoring contractual relationships

An organization must consider everything that is internal and relevant to its mission, strategic objectives, strategic direction, processes, and operations. It needs to understand the influence these considerations could have on its sustainability program and on the results that it intends to achieve.

External Context

The external context includes the broad external operating environment in which the organization operates. Influences and factors in the external context can include the following[5]:

- Cultural, social, political, legal, regulatory, financial, technological, economic, natural, and competitive
- Key drivers and trends exerting positive and negative consequences that affect the objectives
- Relationships with external stakeholders, along with their perceptions, values, and interests

Organizations must assess their external operating environment to determine and characterize the crucial influences and factors that might support or impair their ability to manage the opportunities and threats that are identified. These factors are identified by examining the conditions, entities, and events that determine the associated opportunities and threats, and which may influence the organization's activities and decisions.[6]

Because the external operating environment affects the operations, people involved in supporting the operations have an interest in the information that is created in the characterization of the external environment. The following elements are often considered[7]:

- *Material resources*: Suppliers, real estate, and brokers
- *Human resources*: Labor market, schools, and unions
- *Financial resources*: Banks, investors, granting agencies, and contributions
- *Markets for products and services*: Customers, clients, and users
- *Competitive environment*: Competitors and competitiveness
- *Technology*: Hardware, software, information technology, and production techniques
- *Economic conditions*: Inflation rate, signs of recession, rate of investment, and growth
- *Government oversight*: Regulations, taxes, services, court system, and political process
- *Sociocultural*: Demographics, values, beliefs, education, religion, work ethic, and green movement
- *International*: Exchange rate, competition, selling overseas, and regulations

An organization needs to consider its relationship to each of these elements in terms of their strengths, weaknesses, opportunities, and threats.[8]

Scanning the Organization's Operating Environment

Organizations are connected to an outside world with a constantly changing set of influences and factors. There is often similar change going on within the organization. Organizations should monitor the context to identify, assess, and manage the opportunities and threats associated with change. If the organization is able to analyze these changes that can affect its performance, it can then begin to make decisions and develop a strategy for operating in this uncertain world.

Understanding the context of the organization provides insight into the internal and external influences and factors that impact the way organizations operate at the community level and influence the decisions that will be made by leadership at that location.

Scanning the Internal Environment

Understanding the internal context prepares the organization for managing the opportunities and threats originating within its processes and operations. The internal context is also important to the implementation of any risk management activity by the organization. Most factors associated with the internal context are within the control of the organization or within its sphere of influence.

A methodology widely used in the project management field provides a set of "influences" that are useful for scanning the internal operating environment. This TECOP tool contains the following influences[9]:

- *Technical*: Information and communications technology (hardware and software), internal infrastructure (including its condition), knowledge sharing, research and development, assets, required skill levels and competency of workers, and innovation efforts
- *Economic*: Financial management, financing, cash flow, return on investment, capital reserves, taxes, royalty payments, insurance, and the commercial viability of the organization
- *Cultural*: Demographics, and collective attitudes and behavior characteristics of the organization
- *Organizational*: Capabilities, policies, standards, guidelines, strategies, management systems, structures, and objectives
- *Political*: Governance, internal politics, decision-making systems, stakeholders, roles, and accountabilities

There is another version of TECOP that replaces *cultural* with *commercial*. However, for the purposes of determining the internal context, the cultural category is of great importance for understanding the influence of people on an organization's operations.

When considering the processes and operations of the organization, a SWIFT (i.e., "structured what ifs technique") tool[10] can be used to supplement the characterization of the internal context through the use of the TECOP analysis. SWIFT is a method that enables a facilitator to guide a team through a systematic evaluation using "prompts" to determine how the processes and operations could be affected by opportunities and threats found through the use of the TECOP analysis. Deviations from processes and operations create additional opportunities and threats that must be considered in the internal context.

Scanning the External Environment

The external context is characterized by conducting a broad scan of the external operating environment. A PESTLE (also called PESTEL) analysis tool is used to systematically assess the influences and factors that create opportunities and threats in the external operating environment that are not controlled by the organization. The influences associated with the PESTLE analysis include the following:[11]

- *Political*: Factors include the extent to which governments or political influences are likely to impact or drive global, regional, national, local, and community trends or cultures.
- *Economic*: Factors include global, national, and local trends and drivers; financial markets; credit cycles; economic growth; interest rates; exchange rates; inflation rates; and the cost of capital.
- *Societal*: Factors include culture, health consciousness, demographics, education, population growth, career attitudes, and an emphasis on safety.
- *Technological*: Factors include computing, technology advances or limitations, robotics, automation, technology incentives, the rate of technological change, and research and development.
- *Legal*: Factors include legislative or regulatory issues and sensitivities.
- *Environmental*: Factors include global, regional, and local climate; adverse weather; natural hazards; hazardous waste; and related trends.

Gathering the information to prepare the external context can be time-consuming and difficult. Having an independent analysis conducted by a local team under contract could facilitate the ability to determine the opportunities and threats associated with each of the factors. However, the people working within the organization may have some unique perspectives that a contractor might not be able to develop during the life of the contracted effort. Once there has been some meaningful engagement of the internal stakeholders, it is possible to use their knowledge of the external operating environment to complement this effort.

Adapting to a Changing External Context

Uncertainty in the external context may create a need to make changes in the organizational structure and operating behaviors. An organization in a *certain* external operating environment is managed and controlled differently from an organization operating in an *uncertain* external context. Organizations need to have the correct fit between internal structure and the external environment.

Uncertainty can be expressed in terms of volatility, uncertainty, complexity, and ambiguity[12]:

- *Volatility*: Unexpected or unstable challenge that may be of unknown duration but knowledge about it is often known and available. How volatile is the current situation? What are the nature and dynamics of change, and how can the change affect the organization? What is the nature and speed of the forces of change?
- *Uncertainty*: Despite a lack of other information, an event's basic cause and effect are known; change is possible, but not a given. How much predictability is expected, and which areas of the organization's structure have the least levels of certainty? What factors around the lack of predictability, the prospects for surprise, and the sense of awareness and understanding of factors and events should be analyzed?
- *Complexity*: A situation has many interconnected parts and variables; some information is available or can be predicted, but the volume or nature of that information can be overwhelming to process. How complex is the context, the operating model, and the environment within which the organization operates? What are the combination of forces, the confounding of factors, and the chaos and confusion that can be found in the external context?
- *Ambiguity*: Causal relationships are completely unclear; no precedents exist, facing "unknown unknowns." What level of ambiguity is the organization currently facing, and what is expected in the future? In what areas is the organization facing causal relationships, and how are they likely to affect the organization? What are the key factors around the haziness of reality, potential for misreading, or mixed meaning of conditions and cause and effect?

Volatility, uncertainty, complexity, and ambiguity are referred to as components of complexity theory or the science of complexity. There are a growing number of studies to apply practical approaches to organizations so they can strategize and change.[13]

Finding Opportunities and Threats

Scanning involves the analysis of events, trends, and relationships in the organization's internal and external operating environment. The information obtained from such an effort can assist organization leaders in planning a strategy that can guide the organization in the future. It is important to scan the external operating environment to understand the forces of change, such as volatility, uncertainty, complexity, and ambiguity. Scanning involves both look-ing at information (i.e., viewing) and looking for information (i.e., searching).[14]

Scanning Methods

One method to help understand the external operating environment is to review a wide range of different sources[15]:

- Laws and regulations
- Newspaper
- Electronic media
- Newsletters, magazines, journals, and books
- Reports and presentations
- Interviews

All the information from the strategic planning process should be reviewed. Data from the organization's documented information is quite useful, for both the operations and the supply chain.

Once information from all sources has been reviewed, it is time to start using the PESTLE analysis method. The acronym letters refer to the influences in the external operating environment. Each influence has a number of factors. It does not matter how the factors are allocated to the different influences. The aim of the scanning activity is to identify as many factors as possible.[16]

In order to find out which factors can yield valid opportunities and threats, it is important to have the scanning team learn how to ask effective ques-tions. Questioning is a vital tool that helps obtain data, information, knowl-edge, and wisdom. However, the information obtained is only as good as the questions asked. Questions contribute to the success or failure of finding opportunities and threats as the team seeks to understand a complex and changing external context.[17]

Effective Questioning

The scanning team needs to have a plan to use the questions to catalyze insight, innovation, and action on the part of the organization. Each of the following steps is of critical importance to the discovery of opportunities and threats[18]:

- Assess the current situation with the influence–factor combination.
- Discover the big questions that can help identify and clarify the opportunities and threats.
- Create images of possibilities and scenarios of how the organization may be affected.
- Evolve and create strategic opportunities and threats.

Many organizations use scanning for marketing as well as for strategy. Keep in mind that the methods and outcomes are quite different. Sustainability practitioners are usually keen on finding opportunities, in both the internal and external context. It is important to engage the sustainability team as part of the scanning effort. If there is no sustainability focus in place, the scanning team needs to reach out to people that are involved in the stakeholder engagement process. It is likely that the external stakeholders are already savvy about what is happening in the neighborhood and the community at large. Exchanging information with the stakeholders will help improve the effectiveness of the engagement process (see Chapter 9).

The scanning team should consider the following methods to help improve the effectiveness of their effort[19]:

- Engage in shared conversation.
- Convene and host "learning conversations."
- Include diverse perspectives in all conversations.
- Foster shared meaning.
- Nurture communities of practice.
- Use collaborative technologies.

All these are important elements of effective questioning in a learning organization.

Critical Thinking

After the questioning, the focus shifts to critical thinking. By its nature, critical thinking demands recognition that all questioning[20]

- Stems from a point of view and occurs within a frame of reference
- Proceeds from some purpose—to answer a question
- Relies on concepts and ideas that rest on assumptions
- Has an informational base that must be interpreted
- Draws on basic inferences to make conclusions that have implications and consequences

Each of these dimensions of reasoning is linked to the others. Failure to recognize the links in any of these steps will impact the other steps. The critical thinking process needs to be carefully monitored.[21]

Critical thinking is the practice of analyzing and evaluating thinking with the intention of improving the information that is derived from this thinking. Critical thinking is the purposeful, reflective, reasonable, and self-regulatory process of thinking out possible opportunities and threats from each of the influences and their factors, and determining the evidential, conceptual, methodological, and contextual considerations upon which judgment on these opportunities and threats is based.[22]

The critical thinker[23]

- Raises questions and discusses potential opportunities and threats associated with every influence–factor, being careful to formulate them clearly and precisely
- Gathers and assesses relevant information using critical thinking to interpret the information effectively
- Thinks open-mindedly within the critical thinking realm to recognize and assess assumptions, as well as the implications and consequences from various interpretations and inferences
- Communicates effectively with others to determine the opportunities and threats associated with each influence–factor combination

Critical thinking is analytical, judgmental, and selective. This is in contrast to creative thinking, which is more adaptive to a changing context, especially when that involves escaping from a preordained pattern. Thinking creatively involves generating ideas that are often considered to be unique and plausibly effective. For the purpose of finding the proper opportunities and threats for each influence–factor combination, it is necessary to have some creative thinking capability present on the scanning team or to find that information via a literature search or interviews. The critical thinking will remain the dominant force.[24]

Stability of the External Context

Researchers have noted that a stable external context creates an internal context with a mechanistic bent favoring standard rules, procedures, clear hierarchy of authority, formalization, and centralization. This is in contrast to an external context that is in rapid change, where the internal context appears to be much looser, free-flowing, and adaptive. Rapid change spawns a flexible hierarchy of authority and a decentralized decision-making process. Organization members or employees are encouraged to deal with opportunities and threats by working directly with one another, using collaborative teams, and taking an informal approach to assigning tasks and responsibilities. This helps the organization adapt to continual and sudden changes in the external environment.[25] Organizations that are most successful in uncertain external contexts keep everyone in close touch with influences, factors, opportunities, and threats in the external context. This can hasten the response that may be necessary.

Sustainability and Context

Organizations operating without a focus on sustainability seek to primarily control internal threats. In a financial environment, this is referred to as "internal controls." In an internal context, this is referred to as "operational controls." Scanning of the external operating environment often involves those who are interested and involved in the organization's sustainability efforts. These sustainability efforts focus on effective process and efficient operations within the internal context, but also on searching for opportunities and threats in the external operating environment through their involvement with the stakeholder engagement program. Without a focus on sustainability, the stakeholders are regarded as "interested parties."

Sustainable organizations where there is a high level of stakeholder engagement tend to be more vigilant about changes in the external context.

In some organizations, the information about the internal context can be determined by reviewing existing practices, plans, processes, and procedures. However, many organizations operating at the community level tend to operate informally. Context review team members need to ask questions of the members or employees to determine how they function within the organization. The organization's leader could access the ability of the organization to achieve the identified objectives through some basic level of strategic planning (e.g., strengths, weaknesses, opportunities, and threats). It always helps to see how the organization has dealt with previous failures, incidents, accidents, and emergencies to obtain a more complete picture of an informal internal context.

Changes in either the internal or external context create uncertainty for the organization. Organizations must manage the effects of uncertainty (i.e., the opportunities and threats) to lower the risk to meet their objectives. An increased level of uncertainty is often associated with the inability of internal decision makers to obtain sufficient information about the context factor and the opportunities and threats associated with these factors. Uncertainty increases the risk to meeting the objectives. Opportunities can contribute to the chances that the objectives will be met. Threats are likely to lead to chances that the objectives will not be met.

A complex external operating environment is one in which the organization interacts or is influenced by many diverse PESTLE influences and factors. In contrast, a simple external operating environment leads to an organization that only interacts with or is influenced by a small number of similar factors as determined using the PESTLE analysis.[26] There is also a stable–unstable dimension associated with the external context. In a stable external operating environment, the PESTLE influences and factors remain essentially the same over a period of months or years. During unstable conditions, the PESTLE influences and factors change rapidly. It is very important to constantly monitor the external context under unstable conditions.[27] This is a role that the sustainability team is particularly suited for.

A Path Forward

The context analysis team might consider the following "looking forward" actions:

- Reviewing or creating the risk management and sustainability policies
- Analyzing the organization's resources (e.g., financial capital, human capital, assets, and materials productivity) and knowledge management capability
- Performing a stress test of the decision-making processes, including sense making and knowledge management (see Chapter 3)
- Creating a cohesive set of organizational systems of management for the use of a plan–do–check–act sequence of processes
- Enhancing the organization's contractual and informal relationships that involve adherence to a "supplier code of conduct" as a condition of collaborating at any level with another organization

Organizations will use these actions to protect the operations associated with the internal context and otherwise manage the level of uncertainty to help meet their objectives. To thrive in a world characterized by uncertainty, an organization needs to turn to risk management and sustainability to help take advantage of the opportunities and use them to offset the threats. Treating threats to make them less unacceptable always leads to the creation of additional threats when operating in a system. Sustainability efforts need

BOX 8.1 ESSENTIAL QUESTIONS FOR DETERMINING THE CONTEXT

How do changes that occur in the internal and external operating environments affect the ability of the organization to meet its strategic objectives?

How is sustainability affected by changes in the internal and external context? How does the corporate sustainability manager take this into account when preparing the sustainability report?

Why is it important for a corporation to understand that all its facilities (i.e., organizations) are "different" because of the context even if they have identical products and services?

How can the PESTLE and TECOP tools help standardize the scanning processes at different facilities to provide consistency within the corporate sustainability program?

to be continuously vigilant with respect to opportunities and threats associated with a changing external context. These efforts will also help to engage the external stakeholders to seek their support in this effort. Essential questions for addressing sustainability within the determination of the organization's internal and external context can be found in Box 8.1.

Endnotes

1. AS/NZS, 2013.
2. ISO, 2009.
3. ISO, 2010.
4. ISO, 2009.
5. ISO, 2009.
6. AS/NZS, 2000.
7. Daft, 2013.
8. AS/NZS, 2000.
9. Talbot, 2011a.
10. ISO, 2009a.
11. Talbot, 2011.
12. Bennett and Lemoine, 2014.
13. Serrat, 2009b.
14. Choo, 2001.
15. AS/NZS, 2013.
16. FME, 2013.
17. Serrat, 2009.
18. Serrat, 2009.
19. Serrat, 2009.
20. Serrat, 2011.
21. Serrat, 2011.
22. Serrat, 2011.
23. Serrat, 2011.
24. Serrat, 2011.
25. Daft, 2013.
26. Serrat, 2011.
27. Serrat, 2011.

9

Engagement with Stakeholders and Social License to Operate

Every organization comes face-to-face with a wide variety of different kinds of stakeholders as it begins to develop sustainability by characterizing its internal and external operating environment. A stakeholder is an individual or organization that has an *interest* in a decision or activity of an organization.[1] Interest refers to the actual or potential basis of a claim or demand for something that is owed, or to demand respect for a right. The claim does not need to involve a financial demand or legal right. Sometimes, it is the simple request to have the right to be heard.[2] This interest gives them a "stake" in the organization. However, this relationship is usually not formal or even acknowledged by the stakeholder or the organization. Stakeholders are also referred to as "interested parties." The relevance or significance of an interest is determined by the principles of risk management and sustainability (see Chapter 2).

The three responsibilities of sustainability[3] require the willingness of an organization to operate in a transparent and accountable manner and in compliance with all applicable laws and regulations. It is expected that the organization practices its three responsibilities within its activities, processes, and operations while taking into account the interests of its stakeholder.[4]

Stakeholders

Identification of and engagement with stakeholders is fundamental to the practice of sustainability by an organization. Every organization should determine *who* has an interest in its activities, processes, decisions, products, and services. It needs to determine the consequences of its activities when conducting the internal and external operating environment scanning exercises. While an organization is usually clear about the interests of the owners, members, customers, or other constituents of the organization, there are often people and organizations outside of the control of an organization that have rights, claims, or other specific interests that need to be addressed that are not being addressed. Not all the stakeholders belong to organized groups that have the purpose of representing their interests. In some cases, the "interest" exists whether or not the parties are aware of the mutual interest.

Many organizations are not aware of all their stakeholders, and stakeholders may not be aware of the potential of an organization to affect their interests. To address these conditions, an organization should[5]

- Create a means of identifying its stakeholders and keep that list current
- Recognize and have due regard for the interests and other legal rights of its stakeholders and engage with them regarding their interests
- Assess and take into account the ability of stakeholders to contact, engage with, and influence the organization
- Take into account the relation of its stakeholders' interests to the organization's risk management and sustainability programs, as well as the nature of the stakeholders' relationship with the organization
- Consider the views of stakeholders whose interests are likely to be affected by an activity, process, or decision even if they have no formal role in the governance of the organization or are unaware of these interests

An organization needs to understand the relationship between the various stakeholders' interests that are affected by the organization, while also considering the interests of society as a whole—starting with the community. Although stakeholders are a segment of the larger society, they may have an interest that is not consistent with the expectations of society.[6]

An effective means for an organization to identify its sustainability responsibilities is to become familiar with the practice of scanning its internal and external operating environments while realizing that the stakeholders are the "face" that can be put on the opportunities and threats that have been identified.

Internal Stakeholders

Some stakeholders are involved within the organization. They include members, employees, leaders, and owners. These stakeholders have an interest in the mission and the strategic objectives of the organization. However, this does not mean that all their interests will be the same. Competent, empowered, and engaged people at all levels in the organization are critical to the success of the organization meeting its objectives in an uncertain world. Recognition, empowerment, and enhancement of competence help facilitate the engagement of people in achieving effective processes and efficient operations in the organization.

It is important to remember that internal stakeholders are also connected to external stakeholders through their family and friends in the community. Since everyone belongs to multiple organizations, it is important to have the members or employees assist with the monitoring of the external operating environment and identify which key external stakeholders are known to them.

Another point to remember is that the parent organization is linked to all its organizations through the internal context and reporting structure. This provides some consistency throughout a corporation, but the individual facilities are still all different from one another because of context.

External Stakeholders

All the organizations that are identified in the scan of the external operating environment have people that become external stakeholders. It is important to begin the search for external stakeholders with these people and their organizations. The organization and its parent organization may have a sphere of influence on some of the external stakeholders.

Some organizations consider their value chain organizations as internal stakeholders since there are often contractual arrangements or other factors that create a sphere of influence. However, the level of influence is reduced in the second and third tiers of the value chain.

Identification of Stakeholders

An organization should determine what other organizations or individuals have an interest in its activities, processes, decisions, products, and services. While this is often accomplished as a component of the scanning of the internal and external operating environment (see Chapter 8), extra effort is warranted to enable the organization to understand the relevance of its opportunities and threats and the consequences of these effects of uncertainty on the stakeholders. Putting a "face" on these opportunities and threats helps the organization when it conducts the uncertainty analysis. The identification of stakeholders does not replace the considerations of the broader reach of society in determining norms and expectations of the organization's activities and decisions.

Parent organizations (e.g., corporations) will have global stakeholders, such as the major nongovernmental organizations (NGOs), and multilateral groups, such as the United Nations and the International Labour Organization. Organizations at the community level will deal with local government officials, community government departments, local community board members (e.g., zoning board), neighbors, local advocates of various causes, and people involved in community service. Many people in a community have an "interest" in what an organization is seeking to accomplish with its strategic objectives.

To identify stakeholders beyond its scan of the external operating environment, an organization should consider the following questions[7]:

- To what organizations are there legal obligations?
- Who might be positively or negatively affected by activities, processes, decisions, products, and services?
- Who might express concerns about these activities, processes, decisions, products, and services?
- What involvement have other organizations or individuals had in the past when similar concerns needed to be addressed?
- Typically, who helps the stakeholder organization address specific consequences?
- Which stakeholders can affect the ability of an organization to meet its responsibilities?
- Who would be disadvantaged if excluded from engagement?
- Who in the community or value chain is affected?

Some sustainability practitioners refer to an interest as an "issue." An issue is defined[8] as "a point or matter in question or in dispute, or a point or matter that is not settled and is under discussion or over which there are opposing views or disagreements." When stakeholders have issues rather than interests, the organization has waited too long to engage with them. Learning about an issue signifies that an impasse has been reached and that mediation may be necessary to get past this impasse. By not choosing to deal with stakeholder issues, an organization could jeopardize its social license to operate.

To be sure that an organization has satisfactorily dealt with the identification of its stakeholders, it should consider each of the following tasks[9]:

- Have a process for identifying both its internal and external stakeholders while considering the process for conducting the internal and external operating environment scans
- Recognize and have due regard for the interests, as well as the legal rights, of its stakeholders and engage with them to better understand their interests
- Recognize that some stakeholders can significantly affect the level of uncertainty of an organization and keep it from achieving its strategic objectives
- Assess and take into account the relative ability of stakeholders to contact, engage with, and influence the organization
- Take into account the relation of its stakeholders' interests to the broadest expectations of society and to sustainability, as well as the nature of the stakeholders' relationship with the organization

- Consider the views of stakeholders whose interests are likely to be affected by a decision or activity—this is a prospective view that comes from an effective stakeholder engagement process

Stakeholder identification and engagement are critically important to addressing an organization's three responsibilities of sustainability—environmental stewardship, social well-being, and the sharing of value between organizations in the community.

Stakeholder Engagement

Stakeholder engagement is defined[10] as "activity undertaken to create opportunities of dialogue between an organization and one or more of its stakeholders, with the aim of providing an informed basis for the organization's decisions." This term does not specifically address what constitutes engagement. Some quality management texts have defined[11] *engagement* for members or employees as "the level of connection felt with their employer, as demonstrated by their willingness and ability to help the organization succeed largely by providing discretionary effort on a sustained basis." In this case, engagement refers to the emotional and intellectual commitment to accomplishing the strategic objectives of the organization. Employees feel engaged when they find personal meaning and motivation in their daily efforts and when they receive positive feedback and interpersonal support. An engaged workforce is made possible through the development of trusting relationships, good communication, a safe work environment, empowerment, and performance accountability. A job is only a job until the worker identifies with it and shares in its meaning. It is important that employees know how they contribute to their organization's success and the achievement of the strategic objectives.[12]

External to the organization, engagement can involve similar activities:

- Face-to-face meetings with active listening on both sides
- Involvement taking place over time
- Seeking to monitor the effectiveness of the engagement process

Engagement should be expressed as positive, proactive involvement.[13] When engaged with an organization, individuals must be involved. Interactions occur, information is exchanged, and operational friction is reduced. As in the case with employees, effective engagement with external stakeholders also involves a strong emotional bond to the organization. Unfortunately, few stakeholder engagements ever reach this level. Some organizations are still using surveys to gather information from internal and external stakeholders. The use of surveys does not meet any of the three elements of engagement mentioned above.

Digital media is changing the way we engage with stakeholders. Being digital enables an organization to evaluate how its engagement with external stakeholders may present opportunities and threats. This means understanding how stakeholder behaviors and interests are developing outside of the business, which is crucial to getting ahead of trends that can deliver or destroy value.[14] Digital tools also contribute to rethinking how to use new capabilities to improve how stakeholders are engaged. Data and metrics can focus on delivering insights about stakeholders that in turn drive the ability to understand their interests. Digital tools enable an organization to constantly engage its stakeholders as part of a cyclical dynamic where internal processes and changes in the external operating environment are constantly evolving based on direct inputs from the stakeholders, fostering ongoing engagement and empowerment. Being digital is about using data to make better and faster decisions and moving those decisions lower in the organization. The use of digital tools also speeds the engagement process, enabling fast-moving, stakeholder-facing interactions over time.[15] Many very small organizations are beginning to use digital media in their marketing efforts. They will learn how to extend it to stakeholder engagement.

Another means for improving stakeholder engagement is borrowed from the more traditional voice of the customer (VOC) method.[16] A VOC program is defined as a "channel for acquiring business insight about customers and what is important to them." Capitalizing on customer feedback requires a strategic and ongoing dedication to hearing, active listening, understanding, and acting on the customer's "voice: through a formal program built on[17]

- *Active listening*: A mechanism providing all customers the opportunity to share their compliments and comments about their experiences with an organization's product or service.
- *Pulse monitoring*: A recurring and systematic means of tracking changes in business outcomes, their leading indicators, and the influential drivers by periodically contacting and talking to a statistically representative sample of customers. This is now done digitally instead of using polling.

Listening to customers is now starting to include the remainder of the external stakeholders. This needs to be a central element in the scanning of the external environment as well. Many organizations are beginning to use these VOC techniques with members or employees.[18]

Putting Engagement into Practice

Some actions for engaging members or employees of the organization are[19]

- Recognize that sustainability is embedded in the work that they do.
- Promote collaboration throughout the organization to ensure the organization meets its three responsibilities of sustainability.

- Communicate face-to-face with people to promote understanding of the importance of their individual contributions to sustainability and to assess their satisfaction with the working environment that enables them to realize their commitments.

- Facilitate open discussion and sharing of knowledge and experience between members or employees.

- Recognize and acknowledge people's contribution and learning, as well as the improvement that comes from their efforts.

- Enable their self-evaluation of performance against their personal objectives for sustainability within their control.

- Communicate the results of sustainability and recognize all significant contributions.

When an employee is engaged with an organization, operational friction is reduced while an emotional bond is created. However, employee engagement must be self-motivated rather than simply participating in random sustainability initiatives.

A VOC-type program is much more than the use of an occasional survey. It is recognition of the importance of customers (and employees) to an organization's success and a commitment to include stakeholder perspectives in decisions being made in every part of the operations. Essential questions for addressing sustainability as a critical component for stakeholder engagement and the maintaining of the organization's social license to operate can be found in Box 9.1. It begins with the opportunities and threats gathered in the internal and external context scans, continues with the practical and insightful analysis of stakeholder feedback (i.e., interests), and concludes with the creation of activities guided by the analysis to demonstrably impact the manner in which the organization engages all its stakeholders.[20] This is the way to embed stakeholder engagement into the organization's culture.

Within a sustainability program, employees are engaged, informed, incentivized, motivated, and rewarded for their contributions to the organization's sustainability program. This effort is often operated as a "behavioral change" program with a wide variety of different tools, techniques, frameworks, and approaches. When an organization commits to a VOC-style program, the bolt-on sustainability program has much more difficulty showing how it can add value through its contribution to stakeholder engagement. Sustainability needs to be built into the program.

Social License to Operate

The social license to operate is defined as existing when an organization has the ongoing approval within a community and from other internal and external stakeholders. This can also be expressed as having broad social acceptance or ongoing acceptance.[21] Social license to operate is rooted in the beliefs, perceptions, and opinions of the external stakeholders. While the "license" is intangible to a great extent, an organization recognizes the importance of earning the acceptance or the highest level of approval from its stakeholders. This represents the greater community accepting the organization and its projects into their collective identity.[22]

A social license to operate is usually granted by the stakeholders in the community. Similar organizations might have a social license in one application and not have it for a similar application in the same area or neighborhood. Generally, the more expansive the environmental, social, and economic impacts of an organization and its activities, products, and services, the more difficult it becomes to get the social license.

The critical components of the social license to operate are[23]

- *Legitimacy*: This is based on the norms of the community, which may be legal, social, and cultural, as well as formal or informal, in nature. Organizations must clearly understand the community's "rules of the game." Failure to do so risks rejection. In practice, the social license to operate comes from a program of stakeholder engagement. Nothing should be left to chance. There must be a transparent exchange of information about the activities, decision, product, or service.

- *Credibility*: The capacity to be credible is created by consistently providing accurate and clear information and by complying with any and all commitments made to the community in the process of the stakeholder engagement. Documenting the understandings helps manage expectations and reduces the risk of losing credibility by being perceived as creating a breach of promises. Many experts offer the advice of not making a verbal commitment since, in the absence of a permanent record, there are always areas open to reinterpretation at a later date.

- *Trust*: Trust, or the willingness to be vulnerable to the actions of another, creates a very high-quality relationship and one that takes both time and effort to create. Trust comes from shared experiences gained in the stakeholder engagement process. The challenge for the organization is to go beyond a traditional stakeholder engagement process and create opportunities to collaboratively work together, and generate the shared value within which trust can prosper.

Gaining and maintaining a social license to operate is complex, to say the least. Difficulties arise when organizations are unable or unwilling to make the nominal investment of time to make things work well.[24]

There are a number of common problems associated with obtaining a social license to operate[25]:

- An organization sees the social license to operate as a series of tasks, while the stakeholders and community grant the license on the basis of the quality of the relationship. A cultural mismatch often results in failure.
- The organization mistakes acceptance for approval, cooperation for trust, and technical credibility for social credibility.
- The organization does many of the following actions:
  - Fails to understand the local community's local rules of the game and is unable to establish social legitimacy
  - Delays the start of the stakeholder engagement process when there has been a significant change in its internal and external context
  - Fails to allocate sufficient time for relationship building
  - Undermines its own credibility by failing to give reliable information or failing to deliver on promises made to the key stakeholders or the community
  - Underestimates the time and effort required to gain a social license to operate
  - Overestimates the quality of the relationship with the key stakeholders or the community

While the term *community* is used when talking about social license to operate, it is really about the interests of the key stakeholders, especially when they are in line with the norms of the community and embrace its principles. It is definitely more difficult to create a relationship with stakeholders when their interests are not in line with others' in the community. It is important for organizations to have the ability to use techniques for capacity building at the local level. This is often accomplished with outside consulting help.

There are some cases where it is difficult to achieve a social license to operate. Similar to the practice of sustainability, there is no "one-size-fits-all" activity for either effort. It is always prudent to seek the counsel of people that understand the social structure and let them guide the organization as it performs its analysis of the external operating environment and as it applies this information within the context of its risk assessment process to rank order its opportunities and threats. A risk assessment process can be a valuable asset when an organization is involved in obtaining its social license to operate.

Finally, it is important to remember that the quality of the social license to operate is dynamic and responsive to changes in perceptions regarding the organization and its activities, decisions, and projects. It is also susceptible to outside influences. The organization has to be diligent to work hard to maintain its social license to operate over time once it is obtained.

There are many useful publications dedicated to the social license to operate. Many of them are written for large corporations for use with operations that extract resources from the ground. Be careful that the information used by a specific organization is tailored to that organization and its stakeholders.

Internalizing Stakeholder Interests in an Organization

Organizations of all sizes and types can use risk management[26] to help embed the determination of the internal and external context into the way they operate every day. Here is an example of how the context can be addressed in a sustainability management framework.

Understanding the Needs and Interests of the Stakeholders

The organization determines the internal and external context using the influences determined by the PESTLE and TECOP analysis methods. Context factors must be relevant to its purpose and affect its ability to achieve its responsibly set objectives. Furthermore, the context factors include conditions addressed by the three responsibilities within the organization's sustainability policy. The knowledge gained by understanding the context is considered when establishing and maintaining the organization's sustainability program.

Understanding the Interests of the Stakeholders

The organization must determine the[27]

- Stakeholders that are relevant to the achievement of its responsible objectives
- Relevant interests of the stakeholders
- Need for a social license to operate any portion of its operations, products, or services

The knowledge gained is considered when establishing and maintaining the organization's three-responsibility sustainability program. An effective stakeholder engagement process provides the lookout function, which in turn monitors context so that objectives can remain responsible and the social license to operate can be maintained.

Determining the Scope of the Sustainability Program

The organization determines the scope of the sustainability program by establishing its boundaries and applicability. When determining the scope, the organization considers[28]

- External and internal context
- Opportunities and threats associated with the uncertainty in the context
- Interests of the stakeholders
- Organizational functions and physical boundaries
- Need to obtain and maintain a social license to operate
- Authority and ability to exercise control or influence

The scope should include all activities, processes, products, and services within the organization's control or influence that can pose significant risk to the organization as determined in the risk assessment process. Some organizations share the scope of their sustainability effort with internal stakeholders and many key external stakeholders as well.

Sustainability Framework

In order to continually improve its sustainability performance, an organization may seek to establish, implement, maintain, and continually improve its sustainability framework as informed by the sustainability principles and process outlined in Chapter 6. The organization determines[29]

- How it will satisfy its environmental, social, and economic responsibilities within the sustainability program
- How it will embed this framework into its operations—part of how the organization operates each day
- The influence of its stakeholders and their interests

A simple management system that is tailored to the organization can be prepared with this information. Other information that can be added to the sustainability management system will be presented in the chapters that follow.

Stakeholder Engagement around Sustainability

Sustainability needs to be included within the scanning of the external operating environment and in the uncertainty analysis activity through the use of stakeholder engagement (sometimes referred to as "communication and consultation") because[30]

- The interests of internal and external stakeholders are important to an organization's ability to meet its objectives.

- People will need to take (or not take) actions in order for sustainability to be managed effectively.
- People have some of the knowledge and information upon which effective sustainability management relies (i.e., obtaining the social license to operate).
- Some people might have a right to be informed or consulted about the activities and decisions of the organization.

Stakeholder engagement (i.e., communication and consultation with a two-way dialogue) is a key supporting activity for all processes and decisions within the sustainability and risk management processes.

The organization should develop and implement a plan as to how it will communicate with the external stakeholders. This plan should include[31]

- Engagement of the appropriate external stakeholders and ensuring an effective exchange of information
- External reporting to comply with legal, regulatory, and government requirements
- Providing feedback and reporting on communication and consulting as part of the stakeholder engagement program
- Using engagement communication to build confidence within the organization
- Digitally communicating with stakeholders to make it convenient for them

These planning elements should include processes to consolidate risk management and sustainability information from all available sources.

Remember that engagement, communications, and consultation are processes, not outcomes. They normally take place with stakeholders. The beneficial effects of these processes are based on the transparent exchange of information and persuasion. Decisions are always improved by engagement of stakeholders and keeping them digitally engaged. Stakeholder engagement is what separates sustainability from "business as usual."

Concluding Thoughts

Stakeholder engagement is an essential and helpful component of every sustainability program. Failure to develop stakeholder engagement within the sustainability program can lead to the loss of the organization's social

BOX 9.1   ESSENTIAL QUESTIONS FOR STAKEHOLDER
ENGAGEMENT AND THE SOCIAL LICENSE TO OPERATE

How are the stakeholders identified within the scanning of the
external operating environment?

Why is it so important to engage the external stakeholders as the
means for discussing their interests in the organization's sustainability efforts?

How does true engagement differ from other tools (e.g., surveys)
used to understand the stakeholders' interests?

Why has the social license to operate concept not received greater
use by organizations operating in a community setting?

license to operate. An effective stakeholder engagement process can help to
reduce the effects of uncertainty (i.e., opportunities and threats) caused by
the external operating environment.

It is worth reiterating the importance of a robust stakeholder engagement
effort, especially when there is a good deal of change in the organization's
context. Identifying and engaging stakeholders is how the organization
manages its uncertainty so that it can meet its strategic objectives. With this
understanding, the organization can seek alignment with its stakeholders'
interests while demonstrating how its activities, processes, decisions, products, and services can be of benefit to these people and their organizations.
An organization never deals with all the stakeholders at the same time, but
by following the VOC procedures with the stakeholders, the engagement will
be separate and specific to certain interests. Some of the engagement will be
driven by the stakeholders, while other elements of the engagement will be
initiated and maintained by the organization itself.

Stakeholder engagement needs to be documented as part of the organization's knowledge management effort discussed in Chapter 3. This documentation should be maintained to help improve decision making and help
determine the degree to which sense making needs to be actively scanning
the external operating environment.

Endnotes

1. ISO, 2010.
2. ISO, 2010.
3. Pojasek, 2010.

4. ISO, 2010.
5. ISO, 2010.
6. ISO, 2010.
7. ISO, 2010.
8. Project Management Institute, 2013.
9. ISO, 2010.
10. ISO, 2010.
11. Edmond and Monroe, 2010.
12. Edmond and Monroe, 2010.
13. Conlon, n.d.
14. Dorner and Edelman, 2015.
15. Dorner and Edelman, 2015.
16. Conlon, n.d.
17. Conlon, n.d.
18. Conlon, n.d.
19. ISO, 2014.
20. Conlon, n.d.
21. Thompson and Boutilier, n.d.
22. Thompson and Boutilier, n.d.
23. Thompson and Boutilier, n.d.
24. Thompson and Boutilier, n.d.
25. Thompson and Boutilier, n.d.
26. ISO, 2009.
27. ISO, 2009.
28. ISO, 2009.
29. ISO, 2009.
30. ISO, 2009.
31. ISO, 2009.

10

Organizational Governance and Leadership

Organizational governance is the system by which an organization makes and implements decisions in pursuit of its objectives.[1] Governance systems vary with the size and type of organization and the environmental, economic, political, cultural, and social context associated with its location. Governance from the perspective of an organization is not to be confused with corporate governance, which focuses on governing bodies (e.g., board of directors). Instead, it is a system that has a person or a set of trustees that have the authority and responsibility for realizing the organization's objectives. Organizational governance can be applied in either a formal or informal manner.[2]

Organizational governance is present in every organization as one of its core functions. It provides the framework for decision making at all levels of the organization. The decision making is usually influenced by the leadership of the organization. Governance enables the leadership to take responsibility for the risks associated with its activities and make decisions to embed sustainability into the organization. Responsibility can be extended when the organization and its leaders commit to relationships with other organizations and the community at large.[3]

Effective leadership is critical to the success of the governance system. This is true not only for decision making, but also for effective motivation of the internal stakeholders to make sustainability part of what they engage with every day. Governance seeks to make sustainability part of how the organization operates.

Organizational Governance

Organizational governance provides the framework or rules, relationships, systems, and processes that enable authority to be exercised and controlled within an organization.[4] This includes the manner in which the organization and those people that lead it are held accountable. If the governance is effective, it is more likely that the organization will be able to meet its objectives in an uncertain world.

Governance enables the leaders to direct and control the uncertainty associated with the organization's opportunities and threats.[5] Effective uncertainty management is essential for the organization to understand its risks, modify them as possible, and maximize its chance of achieving its objectives. Risk is always focused on meeting the objectives, which is facilitated by managing the opportunities and threats that contribute to or detract from the objectives. Poor governance increases the uncertainty and leads to problems with the long-term stability of the organization. Governance, management of the uncertainty, and risk are highly interdependent.

It is important to separate the organizational perspective from the focus of "corporate governance." Corporate governance is focused on the fiduciary responsibility of the board of directors.[6]

Governance Principles

There is no "one-size-fits-all" model for governance systems. However, there are some common elements that most leaders believe constitute "good practice" in governance.[7] Governance principles have been developed by the Organisation for Economic Cooperation and Development (OECD)[8] and have been undergoing evolutionary changes to accommodate sustainability programs. The key to success is to have sustainability embedded in the decision-making process at all levels of the organization. This process has the following structural elements: commitment, governance policy, leadership responsibility for governance, and continual improvement.[9] These governance principles overlap with the sustainability principles in the following areas[10]:

- Accountability
- Transparency
- Ethical behavior
- Respect for stakeholder interests
- Respect for legal and other requirements
- Respect for international norms of behavior
- Respect for human rights

The purpose of combining these principles and embedding them in the decision-making process at all levels of an organization is to[11]

- Enhance organizational effectiveness and performance
- Understand and manage uncertainties to minimize the threats and maximize the opportunities
- Enhance the competitiveness of the organization in the local community
- Strengthen confidence of other organizations when a relationship is formed

- Enhance the public reputation of an organization through enhanced transparency and accountability
- Allow organizations to demonstrate how they are monitoring their ethical obligations
- Provide a mechanism for benchmarking the degree of organizational accountability with other organizations in the community
- Assist in the prevention and detection of fraudulent, dishonest, or unethical behavior

Leaders need to integrate these principles with other principles that are mentioned in Chapter 2. There should be a unique set of guiding principles that reflect the unique nature of each organization.

Developing the Governance System

To remain competitive in an uncertain world, organizations must innovate and adapt to their governance practices so they can address the interests of the stakeholders through sustainability and sustainable development. These organizations must also grasp new opportunities to offset the threats that they face. Management of the uncertainty effects must be a key of any governance program.[12]

The structural elements of governance include the following: commitment, governance policy, leadership responsibility for governance, and continual improvement. These elements help implement the organization's commitment to effective governance and address the common elements that constitute good governance practices over the long term. Organizations typically use a formal or informal code of conduct to convey the governance policy.[13]

There are some further operational elements that can improve direction to those with the responsibility for governance within an organization. The elements can include

- Means to identify key governance issues depending on the context of the organization
- Operating procedures for maintaining governance as part of the daily operations of the organization
- Process for dealing with governance breaches and complaints
- Documentation of the governance program
- Internal reporting with links to the organization's strategic development, management of the uncertainty, and organizational objectives

To maintain the governance system over time, the organization must maintain a level of awareness-building activity. The governance must maintain visibility and enable dialogue and communication within the organization.

Governance must be carefully monitored to be able to determine its effectiveness while committing the resources necessary to continually improve. As an organization grows in size, it needs to review the elements of governance on a regular basis. Partnerships and other relationships, formal and informal, are informed of the results of any reviews that take place.[14]

Organizational governance is an important factor in enabling an organization to embed sustainability. Organizational resilience is an outcome of organizational governance.[15]

Organization's Code of Conduct

A code of conduct is a document that can help shape the sustainability culture of an organization. This document sets the standards of behavior expected of all members or employees in an organization.[16] It also helps them deal with ethical dilemmas they experience while serving in the organization. The code of conduct is used by the organization to[17]

- Effectively deal with compliance with relevant laws and other rules
- Help management be more effective in monitoring the culture of the firm and its interactions with a wide range of different stakeholders
- Maintain the integrity and reputation of the organization and its members or employees

By establishing a culture of compliance, the organization can be effective in limiting its liability both in penalties and with its reputation.

A code of conduct should include a clear statement of the organization's commitment to comply with laws and regulations. It must promote an internal culture of fair and ethical behavior and report matters that could prove to be detrimental to the organization's reputation. This commitment can be quite informal in the case of small organizations. A code of conduct is an important element in an organization's risk management program.[18]

The code of conduct applies to all the internal stakeholders. External stakeholders are often asked to comply with the code as a condition of their engagement with the organization, particularly when there are matters that could impact the organization's compliance status or reputation. Many of the operational elements in the governance program should also be applied to the use of the code of conduct. Typically, the code of conduct covers environmental, social, and economic responsibilities to maintain a clear linkage with the sustainability program.[19]

In the true meaning of embedded sustainability, the governance, code of conduct, and three responsibilities of sustainability (environment, social, and economic) are linked both within the organization and with the engagement with stakeholders. Sustainability is informed by the risk management process, the management of the uncertainty (i.e., the opportunities and threats), and a number of principles and processes. People and processes are vital to the long-term viability of an organization. Without all these elements present and aligned with each other, it would be difficult to fully embed sustainability, governance, and the code of conduct into a uniform decision-making framework used at every level of the organization.

Leadership

Effective leadership is critical to the success of the governance system and the ability of the organization to meet its strategic objectives. Leaders need to address the following actions[20]:

- Develop the mission, vision, values, and ethics while acting as a role model
- Define, monitor, review, and drive the continual improvement of the organization's processes and operations
- Engage directly with all stakeholders
- Reinforce a culture of operational excellence with the organization's members or employees
- Ensure that the organization is resilient in order to manage change effectively

Some of the personal traits of leaders are as follows[21]:

- Ability to communicate a clear direction for the organization with a defined strategic focus
- Understand the management of uncertainty in order to enable achievement of the strategic objectives of the organization
- Ability to be flexible in adapting and realigning the direction of the organization in light of changes in the external operating environment
- Recognize that sustainable success relies on continual improvement, innovation, and learning
- Ability to use knowledge management and sense making to improve the reliability of decisions at all levels of the organization
- Inspire while creating a culture of engagement, ownership, transparency, and accountability
- Be the role model for integrity, sustainability, environmental stewardship, social well-being, and shared value—both internally and in the community—in a manner that develops, enhances, and protects the organization's reputation

Leaders are responsible for creating the mandate and commitment to sustainability.

Mandate and Commitment

Leaders of organizations need to demonstrate their commitment to governance with the following actions[22]:

- Ensuring that the risk management and sustainability policies are established in accordance with the organization's objectives, context, and strategic direction
- Ensuring that all the structural elements in this book are addressed with processes in all the operations of the organization
- Taking accountability for the effectiveness of the processes used and the efficiency of the operations associated with the organization's products and services
- Ensuring that risk management and sustainability are embedded in the process approach to the operations
- Communicating the importance of risk management and the responsibilities associated with sustainability to all stakeholders within the engagement process
- Ensuring that risk management and sustainability effectively contribute to enabling the organization to meet its objectives

The purpose of the mandate is to ensure that the organization clearly understands the benefits of risk management and sustainability and will embrace the change involved by embedding these practices into the manner in which the organization operates each day. It is necessary for the leader to be deeply involved in these programs in order to establish credibility with the external stakeholders.

Each organization should consider the following questions when establishing its mandate and commitment to risk management and sustainability[23]:

- How is sustainability embedded in the objectives of the organization, and what goals and action plans are in place to help ensure success?
- Is the leadership clear about the significant opportunities and threats associated with the internal and external context and willing to engage with all stakeholders to make sure the organization can effectively manage the effects of uncertainty?
- Does the leader need to make changes to the prevailing risk attitude[24] that will facilitate the embedding of sustainability in the organization's activities, processes, and decision making?

- Does the form of the policy that provides for the embedding of risk management and sustainability support the other policies that direct the way the organization is operated?

Risk management and sustainability need to be viewed as central to the "process focus" that describes the organization's operations.

Leaders must reinforce the organization's commitment and mandate through the following actions[25]:

- Make sure the risk management and sustainability objectives are carefully linked to the objectives derived from the mission statement
- Make it clear that risk management and sustainability is about effectively delivering the organization's strategic objectives with a uniform program of top-down objectives and a feedback loop of employee goals and action plans (see Chapter 7)
- Ensure the risk management and sustainability activities required by the mandate are fully integrated into the governance and the organization's processes at the strategic, tactical, and operational levels
- Make the commitment to make the necessary resources available to assist those accountable and responsible for managing risk and sustainability
- Require regular monitoring and measurement on the risk assessment and sustainability processes to ensure that they remain appropriate and effective
- Monitor to ensure that the organization has a current and comprehensive understanding of its opportunities and threats and that they are within the determined risk criteria
- Initiate corrective actions when threats are deemed to be unacceptable when they are not able to be offset by the opportunities
- Lead by example
- Review the commitment to the mandate as time, events, decisions, and the external operating environment conditions create change

Meeting the organization's objectives is considered to be essential to the organization. Operating in an uncertain world makes it more important that the leadership is committed to risk management and sustainability, and that these practices are explicitly recognized in everything the organization does.

Sustainability Policy

Once the mandate is in place, it is important for leaders to make the mandate explicit through the issuance of a risk management and sustainability policy. The purpose of this policy is to provide an effective means to clearly express the organization's intentions and requirements. The effectiveness rests on the expression of the organization's motivation for managing risk and sustainability and explicitly establishing what is required and by whom. Leaders must set the proper tone by emphasizing the necessity of using risk management and sustainability to help manage the opportunities and threats that define the uncertainty within which the organization operates. It goes without saying that short policies have a tendency to be more effective. The policy should be succinct, as well as short in length.

Effective communication of the policy also requires confirmation that the information has been understood by the stakeholders inside and external to the organization. Stakeholders must truly believe that the policy reflects the organization's intent to embrace risk management and sustainability as part of how it operates. To secure appropriate buy-in, the leader will need to produce tangible objective evidence that the organization has adopted changes in the way it has operated prior to the adoption of the risk management and sustainability policy, and that those changes are able to persist over time. The organization's reputation and the credibility of its risk management and sustainability management efforts are dependent on its success in meeting the commitments embodied in the policy.

A risk management and sustainability policy is established, reviewed, and maintained by the leaders of an organization. Generally, it must meet the following requirements[26]:

- Be appropriate to the purpose and context of the organization
- Provide a framework for setting and reviewing objectives affected by the risk management and sustainability contributions
- Include a commitment to satisfy all applicable legal and other requirements
- Include a commitment to effective processes, efficient operations, and efficacious strategy
- Include a commitment to continual improvement, innovation, and learning
- Be maintained as documented information
- Be communicated within the organization
- Be a key part of the engagement with the external stakeholders

All reviews of the risk management and sustainability processes begin with the examination of the sustainability policy. It is the keystone element of all activities, processes, decisions, products, and services associated with the organization.

Organizational Roles, Responsibilities, and Authorities

As required by the organizational governance, leaders must ensure that the responsibilities and authorities for relevant roles are assigned, communicated, and understood within the organization.[27] Leaders need to assign responsibility for the following:

- Ensuring the risk management and sustainability processes are established with reference or benchmarked to "best practice" for the organization
- Ensuring that the processes are effectively delivering their intended outcomes
- Reporting on the performance of the risk management and sustainability processes, on opportunities for improvement, and on the need for change or innovation
- Ensuring the promotion of risk management and sustainability throughout the organization
- Ensuring the integrity of the risk management and sustainability processes when changes are planned and implemented in the organization

It is important to have a process focus for the management of risk and sustainability that goes beyond the preparation of a sustainability report. The focus must be on the identification of processes and operations that must be controlled to maintain and improve the organization's risk management and sustainability performance.

Organizational Strategy

At the organization level, it is important to set a strategy to meet the strategic objectives typically in a 5- to 10-year time frame. The effects of uncertainty (i.e., opportunities and threats) must be managed in order for the strategy to be executed in the most effective manner. With this in mind, a strategic plan provides a set of processes that are executed within the organization in order to meet the overarching objectives in an uncertain world. The focus is on developing strategies that will drive the organization in achieving the objectives that have been derived from the mission of the organization.

Organizations exist to create benefits for their internal and external stakeholders. The mission and the strategic objectives define these benefits. Products and services are created within the operations, thereby representing the tactics of the organization. Organizational strategy helps the governance be accountable. This strategy also helps the leaders make decisions that support its execution. By binding together the governance and the

FIGURE 10.1
Strategy links governance and leadership in an organization. (Adapted from Serrat, O., From strategy to practice, *Knowledge Solutions*, Asian Development Bank, Manila, 2009c, http://digitalcommons.ilr.cornell.edu/cgi/viewcontent.cgi?article=1154&context=intl, retrieved June 29, 2015.)

leadership, the organization is prepared for execution (Figure 10.1). Most strategic failures are commonly failures of execution.[28]

A strengths, weaknesses, opportunities, and threats (SWOT) analysis helps organizations use their strength to take advantage of the opportunities and overcome the threats. This analysis also helps organizations in minimizing the weaknesses that are subject to the threats.[29]

Strategic planning must account for the effects of uncertainty and how they affect the future. A scenario is an internally consistent view of the future. Scenario planning is the process of generating and analyzing a set of different futures.[30] The results from building scenarios are not an accurate prediction of the future; rather, they provide better thinking about the future. Scenarios are used in strategic planning to provide a context for decisions. As events and incidents unfold in the external operating environment, it is necessary to continue to review whether plans fit the realities of the PESTLE analysis results. If they do not, the planning team needs to revise its scenario analysis.[31]

Execution of a strategy is a process. It is not an action or a step, and it depends on more people than the strategy formulation. An organization navigates its strategy by focusing on the core building blocks (Figure 10.2).[32] It is best to move in small steps while maintaining a balance between strategizing and learning modes of thinking. Learning adheres to the same principles as the process of evolution.[33] Given the effects of uncertainty on the objectives, there is little alternative to adaptation. Only through action can an organization and its stakeholders participate and gather the experience that both sparks and is informed by the process of learning.[34]

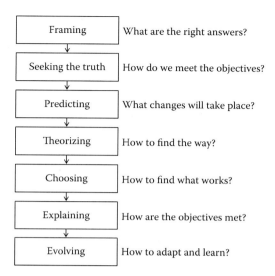

Framing	What are the right answers?
Seeking the truth	How do we meet the objectives?
Predicting	What changes will take place?
Theorizing	How to find the way?
Choosing	How to find what works?
Explaining	How are the objectives met?
Evolving	How to adapt and learn?

FIGURE 10.2
The building blocks of organizational strategy help organizations meet their objectives. (Adapted from Serrat, O., Asking effective questions, Knowledge Solutions, Asian Development Bank, Manila, 2009, http://adb.org/sites/default/files/pub/2009/asking-effective-questions. pdf; Bradley, C., et al., *McKinsey Quarterly*, October 2013, http://www.mckinsey.com.)

BOX 10.1   ESSENTIAL QUESTIONS FOR GOVERNANCE AND LEADERS TO CREATE AN EFFICACIOUS STRATEGY

How does an organization go about creating a "system" by which it makes and implements decisions that will help it meet its strategic objectives?

In what way would a decision-making framework resemble the knowledge management process developed by Olivier Serrat (Asian Development Bank)?

Why is leadership a process rather than a personal trait or characteristic of a person who is a leader?

How are leadership and management alike and different from each other?

A risk assessment is used to prioritize the opportunities and threats in order to ensure action that creates an efficacious strategy and strategic decisions that will enable the organization to meet its objectives. When there are frequent changes in the external operating environment, a risk assessment of the strategic options is required. It will facilitate the analysis of the stakeholder interests, customer requirements for products and services, competencies of the members or employees, and the PESTLE and TECOP analyses.[35] Strategies that are not executable are of no use. Essential questions for governance and leaders to address sustainability within the organization's strategy creation can be found in Box 10.1.

Endnotes

1. ISO, 2010.
2. ISO, 2010.
3. ISO, 2010.
4. ISO, 2013.
5. ISO, 2013.
6. Serrat, 2011a.
7. AS, 2003.
8. OECD, 2004.
9. AS, 2003.
10. ISO, 2010.
11. AS, 2003.
12. OECD, 2004.
13. AS, 2003a.
14. AS, 2003a.
15. BS, 2014.
16. IFAC, 2007.
17. AS, 2003a.
18. AS, 2003a.
19. AS, 2003b.
20. EFQM, 2010.
21. EFQM, 2010.
22. ISO, 2014a.
23. ISO, 2013.
24. ISO, 2009.
25. ISO, 2013.
26. ISO, 2014a.
27. ISO, 2014a.
28. Serrat, 2009c.
29. Kiptoo and Mwirigi, 2014.
30. Serrat, 2009c.
31. Serrat, 2009c.
32. Bradley et al., 2013.
33. Serrat, 2009c.
34. Serrat, 2009c.
35. Hopkin, 2012.

11

Uncertainty Assessment of Opportunities and Threats

What makes risk management unique among other types of management is that it specifically addresses the effect of uncertainty on objectives. Risk can only be assessed or successfully managed if the nature and source of that uncertainty are understood. Uncertainty in the internal and external operating environment of the organization creates a large number of opportunities and threats that must be prioritized for further consideration and management. Stakeholders may contribute additional opportunities and threats during the engagement process with the organization. Leaders encounter opportunities and threats in their decision making. In order to plan for risk management and sustainability, a risk assessment process can be used to determine the significant opportunities and threats identified in these activities.

Risk in an organization is associated with the prospects of meeting the strategic objectives. By managing the effects of uncertainty, there is a much more direct path between risk and the objectives. Risk management represents the "coordinated activities to direct and control an organization with regard to risk."[1] Significant opportunities contribute positively to the objectives, while significant threats contribute negatively to the objectives. Risk management works best when opportunities offset threats, significant opportunities can be embraced, and significant threats can be controlled or otherwise mitigated. All these processes operate in an organization and are not the same kinds of risk common to other applications and fields. For now, the focus is on the opportunities and threats.

Uncertainty Assessment Process

The purpose of the uncertainty assessment process is to provide what is referred to as evidence-based information and analysis that assists in making informed decisions on how to manage the uncertainty associated with opportunities and threats.[2] Risk assessment methods have been slow to adapt to this need to handle positive and negative effects at the same time in the same analysis. This is why we now refer to the uncertainty assessment process rather than the risk assessment method (Figure 11.1).

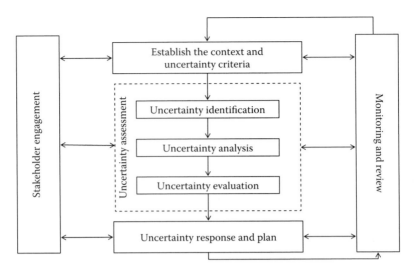

FIGURE 11.1
Structure for uncertainty analysis. (From ISO, Risk management—Principles and guidelines, ISO 31000, ISO, Geneva, 2009.)

Defining the Criteria

The uncertainty assessment process depends on the definition of criteria that will be used to provide terms of reference against which the significance of an opportunity or threat is evaluated.[3] Risk criteria provide the means to compare the opportunities and threats and determine their significance. Each organization creates its own procedure for determining the significant opportunities and threats. The procedure needs to be accepted by the leaders. Once established, the risk criteria are applied throughout the organization whenever there are new opportunities and threats or when opportunities or threats are duly managed.[4] Risk criteria are often discussed within the stakeholder engagement process.

Risk criteria are generated in the following steps[5]:

1. *List outcomes for each strategic objective*: The desired outcomes for each of the organization's objectives should be identified. In the uncertainty analysis, these outcomes and the way they are measured become the method of expressing the "consequence."

2. *Select measures and scales for each outcome to characterize consequences*: For each outcome, the organization needs to consider qualitative or quantitative measures that reflect the degree of success in achieving the objective, along with an appropriate scale on which to express each measure. Consequences can be beneficial (objectives) or detrimental (threats). It is typical to have five levels for characterizing the consequences.

3. *Decide how likelihood will be expressed*: Likelihood can be measured in terms of probabilities (i.e., a value between 0 and 1) or frequencies (i.e., the number of occurrences over a unit of time, or by using descriptive scales). The final case is appropriate where reliance must be placed on judgment and experience. Likelihood scales relate to the likelihood of experiencing the consequences across a relevant time frame, the lifetime of a person, an expected life of an asset, the duration of a project, or government decisions affecting many generations. These ranges must be capable of expressing and distinguishing consequences that are almost inevitable and highly likely from those that are expected to occur infrequently or are improbable.

4. *Decide how consequences and likelihood will be combined to derive the level of uncertainty*: The simplest way to combine consequences and likelihoods in qualitative risk analysis is to use a table to indicate the level of uncertainty that the organization decides should correspond to each combination. Such a table is very important to the organization since it will determine how all opportunities and threats are evaluated. The table should be vetted within the stakeholder engagement and approved by the leaders.

5. *Determine how the level of uncertainty will be expressed*: Simple labels such as *high*, *medium*, and *low* can be used to express the level of uncertainty, or a numerical scale can be used. To avoid confusion, it is important that the terms (or numbers) used to describe the level of uncertainty are different from those used to describe levels of consequences or likelihood.

6. *Decide on the rules for determining the significance of opportunities and threats*: In most cases, the level of significance will be linked to both an authority to accept and some indication of the need for embracing opportunities or avoiding threats.

Establish the Uncertainty Context

The term *context* is used in different way. In an organization's system of management, context refers to an understanding of the internal and external context of the organization. Understanding the context helps identify many of the opportunities and threats. In uncertainty analysis, the purpose of the context is to reveal the sources of uncertainty that relate to the relevant objectives and the particular decision that the uncertainty analysis is seeking to make regarding the uncertainty response and plan.[6] The following questions should be asked when establishing the uncertainty analysis context[7]:

- What is the risk assessment and sustainability policy?
- What are the major consequences and outcomes expected?

- What are the financial implications to the organization?
- What are the significant factors in the organization's internal and external operating environment?
- What are the related opportunities and threats?
- Who are the stakeholders associated with the opportunity and threats?
- What problems were identified in previous uncertainty analyses?
- What uncertainty criteria should be established for the analysis?
- What is the best way of structuring the opportunity and threat identification task?

Attention to these and other relevant factors should help ensure that the uncertainty analysis approach adopted is appropriate to the organization's needs and to the risk associated with the achievement of the objectives.

Stakeholder Engagement

Communication and consultation for the uncertainty analysis take place within the stakeholder engagement process. In this respect, stakeholders should be both internal and external to the organization. Managing uncertainty is important to people because

- The interests of stakeholders are reflected in the establishment or responsible objectives for the organization
- People will need to take (or not take) particular actions in order for the opportunities and threats to be managed effectively
- People have considerable knowledge and information on which effective uncertainty analysis relies
- Some people might have a right to be informed or consulted

Stakeholder engagement is a key supporting activity for all parts of the risk assessment and sustainability program.

Uncertainty Identification Process

Uncertainty identification is a process used to compile a list of opportunities and threats that might contribute to or detract from the achievement of the organization's objectives. For each opportunity or threat, it is important to know the what, where, when, why, and how something could happen and the range of possible outcomes that affect the organization's objectives. It is important to link the factors and the specific sources of the opportunities and threats, along with their consequences (positive or negative). Events and leader's decisions can also cause uncertainty and should be documented.

The information on the opportunities and threats should be as complete as possible. It is normally compiled in a risk or uncertainty register. Failure to identify an opportunity or threat can result in an unexpected threat or a missed opportunity.

Uncertainty[8]

- Is a consequence of underlying sociological, psychological, and cultural factors associated with human behavior
- Is produced by natural processes that are characterized by inherent variability, for example, weather
- Arises from incomplete or inaccurate information, for example, due to missing, misinterpreted, unreliable, internally contradictory, or inaccessible data
- Changes over time, for example, due to competition, trends, new information, or changes in underlying factors
- Is produced by the perception of uncertainty, which may vary between parts of the organization and its stakeholders

An opportunity or threat is generally associated with the following components[9]:

- A source of the uncertainty
- An event, decision, or change in context
- A consequence, outcome, or impact on stakeholders of organization assets
- A cause (what and why), usually a string of direct and underlying causes for the presence of an event, decision, or change in context
- Operational controls and levels of effectiveness
- The timing of the uncertainty and its likelihood

It is important to identify the sources of uncertainty and the interests of the internal and external stakeholders. An uncertainty analysis may concentrate on one or many possible areas of consequences (positive or negative) that may affect the organization. Information on the uncertainty must be sufficient enough to allow understanding of the likelihood and the consequences of the opportunity or threat. Existing information sources (e.g., stakeholder engagement and external context observations) need to be accessed and new data sources developed. Some uncertainties are particularly difficult to observe.

A number of questions need to be asked for each opportunity and threat:

- What is the *source* of each opportunity and threat?

- What might happen that could
  - Increase or decrease the effective achievement of the organization's objectives?
  - Make the achievement of the organization's objectives more or less efficient (financial, people, and time)?
  - Cause stakeholders to take action that may influence the achievement of the objective?
  - Produce additional benefits?
- What would the effect on objectives be?
- When, where, why, and how are these opportunities and threats likely to occur?
- Who might be involved or impacted?
- What controls presently exist to respond to the opportunities and threats?
- What could cause the control not to have the desired effect on the risk associated with the uncertainty?

As with all steps of the uncertainty analysis process, opportunity and threat identification should be iterative and repeated until a level of resolution and accuracy is obtained that is sufficient for the decisions concerned.

The naming of the opportunities and threats is crucial to effective uncertainty analysis and evaluation. Meaningful names should be given to the causes of opportunities, threats, causes, or initiating events to help talk about them. It is important to know how they might trigger an event and the consequences that might be associated with that event. Ideally, an opportunity or a threat should be identified in the following terms: (*something happens*) *leading to* (*outcomes expressed in terms of its consequences with regard to achieving the organization's objectives—positive and negative*). The international risk management and uncertainty analysis does not recommend the use of a risk classification system.

Some organizations use a risk register to manage information on the opportunities and threats. However, the document or database needs to be kept up to date. Risk and uncertainty change all the time. New opportunities and threats appear and old ones disappear.

A risk matrix is also used for ranking and displaying the opportunities and threats by defining ranges for consequences and likelihood. Risk matrices should also be used with great care. It is fine if the matrix is used to crudely compare and rank opportunities and threats. It is important to note that a simple matrix can lead to the false impression of certainty about a consequence or its likelihood, or both.

The documented information from an opportunity and threat uncertainty determination should include the following[10]:

- Approaches and methods used
- Scope of the identification process

- Participants in the uncertainty identification and the information sources consulted
- Risk register with information on the opportunities and threats

Uncertainty identification establishes the information necessary for the uncertainty analysis.

Uncertainty Analysis

Uncertainty analysis involves developing an understanding of the opportunities and threats. This provides an input to uncertainty evaluation and to decisions on whether threats need to be treated and opportunities need to be embraced and developed. Uncertainty analysis also helps the organization make decisions regarding its options to deal with different types and levels of uncertainty and its effect on the level of risk. The results of the uncertainty analysis involve the preparation of a risk map (Figure 11.2).

Uncertainty analysis is defined as a process to comprehend the nature of the opportunity and threat and to determine the level of risk.[11]

Consequence is defined as[12] the outcome of the opportunities and threats, or an event or decision affecting the meeting of the organization's objectives. A means for qualitatively measuring consequence can be found in Figure 11.3.

An event (i.e., occurrence or change of the context of the organization) can lead to a full range on consequences, each of which can be certain or uncertain and have positive or negative effects on objectives. Consequences can be expressed quantitatively or qualitatively.

Likelihood is defined as[13] the chance of something happening. In uncertainty assessment, this term is used to refer to the chance of something happening, whether defined, measured, or determined objectively or subjectively, quantitatively or qualitatively, and described using general terms or expressed mathematically. A means for qualitatively measuring likelihood can be found in Figure 11.4.

Here are some key questions to ask during uncertainty analysis:

- What is in place that may prevent, detect, or lower the consequences or likelihood of the threats?
- What systems may enhance or increase the consequences or likelihood of opportunities or beneficial events?
- What are the consequences or range of consequences if they do occur?
- What is the likelihood or range of likelihoods of the opportunities or threats happening?
- What factors might increase or decrease the likelihood or the consequences?

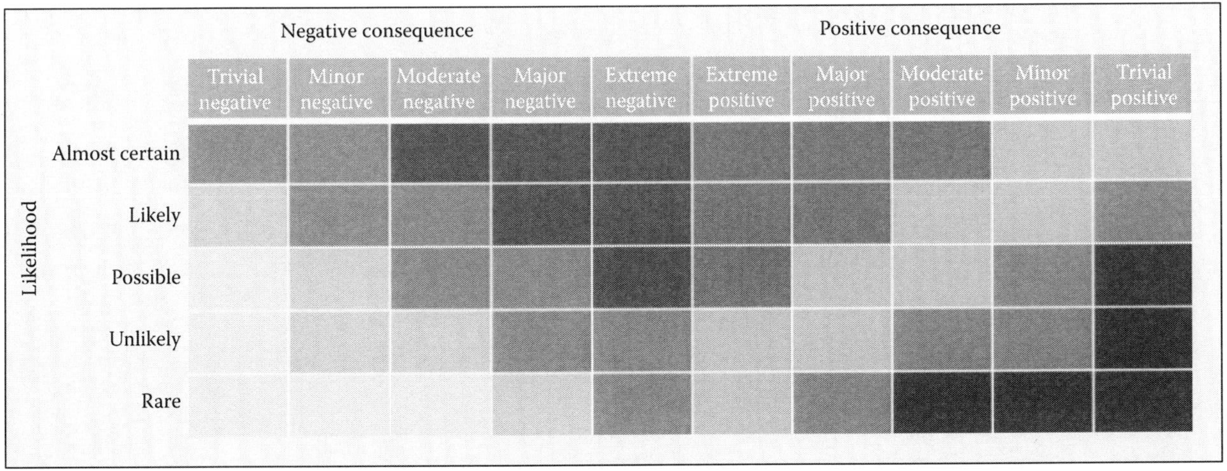

FIGURE 11.2
Risk matrix for the uncertainty analysis. (From Knight, K., Risk management—A journey not a destination, 2010, http://en.mgubs.ru/images/Image/A%20Journey%20Not%20A%20Destination.pdf.pdf.)

Negative consequence	
Extremely negative	Most objectives cannot be achieved
Major negative	Some important objectives cannot be achieved
Moderate negative	Some objectives affected
Minor negative	Minor effects that are easily remedied
Trivial negative	Negligible negative impact upon objectives
Positive consequence	
Extremely positive	Most objectives should be achieved
Major positive	Some unimportant objectives cannot be achieved
Moderate positive	Some objectives affected in a positive sense
Minor positive	Minor improvement to chance of meeting objectives
Trivial positive	Negligible positive impact upon objectives

FIGURE 11.3
Determine the consequence of the uncertainty. (From Knight, K., Risk management—A journey not a destination, 2010, http://en.mgubs.ru/images/Image/A%20Journey%20Not%20A%20 Destination.pdf.pdf.)

Likelihood	
Almost certain	Consequence occurs often
Likely	Consequence occurs several times a year
Possible	Consequence occurs once every several years
Unlikely	Consequence occurs after decades
Rare	Heard of consequence occurring elsewhere

Qualitative analysis often used when the level of risk does not justify the time and resources needed to perform a numerical analysis. This is wellsuited for use in the analysis of opportunities and threats in the organization's context.

Semiquantitative analysis uses numerical ratings or more detailed descriptions for likelihood and consequence.

Quantitative analysis is used when the likelihood of occurrence and the consequence can be reliably quantified.

FIGURE 11.4
Determine the likelihood of the uncertainty. (From Knight, K., Risk management—A journey not a destination, 2010, http://en.mgubs.ru/images/Image/A%20Journey%20Not%20A%20 Destination.pdf.pdf.)

- What additional factors may need to be considered beyond which the analysis does not hold true?
- What are the limitations of the uncertainty analysis and the assumptions that are made?
- How confident are you in your judgment or research specifically in relation to high consequences and low-likelihood opportunities and threats?

- What drives variability, volatility, or uncertainty?
- Is the logic behind the uncertainty analysis method sound?
- For quantitative analysis, what statistical methods may be used to understand the effect of uncertainty and variability?

Where there is a high level of uncertainty remaining following the analysis, it may be appropriate to flag this and review the work at some future date and in light of experience.

Qualitative uncertainty analysis is often used when the level of uncertainty does not justify the time and resources needed to perform a numerical analysis, where the numerical data is inadequate for a quantitative analysis, or to perform an initial screening of uncertainty prior to the detailed analysis.[14] The value of qualitative analysis of uncertainty is maximized when the uncertainty is being determined and shared across a large number of people with different backgrounds and interests. The collection of many ideas within the organization and throughout stakeholder engagement often improves the usefulness of the uncertainty analysis.

Semiquantitative uncertainty analysis uses numerical rankings (e.g., high, medium, and low) or more detailed descriptions for likelihood and consequences. This analysis can move to quantitative analysis when the likelihood of occurrence and the consequences can be quantified.

Uncertainty Evaluation

The purpose of uncertainty evaluation is to assist in making decisions on the uncertainty response based on the outcomes of the uncertainty *analysis*. This enables an organization to determine which opportunities and threats need to be responded to. Uncertainty evaluation compares the level of uncertainty found during the analysis process with the uncertainty criteria established when the context was considered.

Decisions for responding to uncertainty should take account of the wider context of the uncertainty and include consideration of the tolerance of opportunities and threats borne by stakeholders. Decisions should be made with full consideration of legal, regulatory, and other requirements. In some cases, the uncertainty evaluation can lead to conducting more detailed analysis. The uncertainty analysis may also lead to a decision to respond to the opportunity or threat by only maintaining existing operating controls. Uncertainty analysis is influenced by the organization's risk attitude and the risk criteria.[15] Based on this analysis, the need to avoid threats and embrace opportunities can be considered.

The generic strategy for uncertainty evaluation consists of four steps[16]:

1. *Eliminate uncertainty*: Seek to remove threats or offset the threats with opportunities, remembering that opportunities have risk associated with them should they not be able to be exploited successfully.

2. *Allocate ownership*: Seek to either transfer uncertainty to a third party to share in the effect of a loss for a fee, or transfer ownership of an opportunity to a third party that can maximize the benefit with a payment to the organization.

3. *Modify exposure*: In the management of threats, this is referred to as mitigation, an attempt to make the likelihood or consequences smaller. When an opportunity is successfully implemented, it increases its likelihood and consequences on the upside, thereby avoiding the need to mitigate a threat.

4. *Include in the baseline*: Opportunities and threats that are not significant as a result of the uncertainty analysis can be put on a watch list to make sure that nothing is changing to make them more significant uncertainties. It is important to consider multiple opportunities and threats that alone would end up in the baseline, but when clustered may need some other form of response.

These four uncertainty evaluation categories can be looked at from the perspective of threats[17]:

- *Avoid*: Seek to remove threats to lower or eliminate uncertainty
- *Transfer*: Allocate ownership to enable effective management of a threat, often using an insurance company for this purpose
- *Mitigate*: Reduce the likelihood or consequence of the threat below an acceptable threshold
- *Accept*: Recognize residual risks associated with uncertainty and devise ways to control or monitor them

An uncertainty may be tolerated if the consequence or likelihood of that uncertainty is consistent with the established uncertainty criteria and their threshold of what the organization would consider to be an unacceptable exposure. The ability of the organization to absorb an incident will, to a large degree, depend on its size and financial conditions.

Opportunity uncertainty response takes into consideration how to act in order to improve the likelihood and impact of an *opportunity*. When speaking about opportunity response options, there are four major categories of controls that can be used[18]:

- *Exploit* identified opportunities, removing uncertainty by seeking to make the opportunity succeed.
- *Enhance* means increasing its positive likelihood or consequence to maximize the benefit of the opportunity.
- *Share* opportunities by passing ownership to a third party best able to manage the opportunity and maximize the chance of it happening.

- *Ignore* opportunities included in the baseline, adopting a reactive approach without taking explicit actions.

Uncertainty Response

Uncertainty response involves selecting one or more significant opportunities and threats and exploiting the opportunities while avoiding the threats. In some organizations, avoiding the threats is referred to as "risk treatment." Most organizations are still focused only on threats. In other organizations, leaders do not always view opportunities as being able to offset threats. There is a risk associated with the funding of efforts to take advantage of an opportunity. What if the opportunity is not realized despite an effort to take advantage of it? Without the implementation of the opportunity, it would be difficult to monetize the benefit associated with the opportunity. The constraint to developing opportunities as an uncertainty response is solely a function on an organization's experience and comfort developing opportunities.

Several criteria have been defined to assess the effectiveness of these uncertainty responses[19]:

- *Appropriate*: Correct level of response based on the size of the uncertainty.
- *Affordable*: The cost-effectiveness of responses should be determined.
- *Actionable*: Need to identify action timeline since some uncertainty requires immediate attention, while others can wait.
- *Agreed*: The consensus and commitment of the stakeholders should be obtained before creating the response.
- *Allocated and accepted*: Each response should be owned and accepted to ensure a single point of responsibility and accountability for implementing the response.

These criteria work well with both threats and opportunities.

Monitoring and Review of Process

Monitoring and review is an essential step in the process of managing uncertainty. It is essential to monitor and review the effectiveness and efficiency of controls or execution and the appropriateness of uncertainty responses selected. In this way, it is possible to determine if the organization's resources were put to the best use and the uncertainty was reduced to enable the risks to be determined with respect to the organization meeting its objectives.

The organization's monitoring and review processes should include[20]

- Ensuring that controls are effective and efficient in both design and operation

- Obtaining further information to improve uncertainty analysis
- Analyzing and learning lessons from events (including near misses), changes, trends, successes, and failures
- Detecting changes in the external and internal context, including changes to uncertainty criteria and the risk itself—this may require revision of uncertainty responses and priorities
- Identifying emerging risks that lead to more uncertainty

Progress in implementing the uncertainty response should be measured. The results can be incorporated into the organization's performance program and the internal and external reporting activities in the risk management and sustainability programs.

Uncertainty Management Plan

An uncertainty management plan provides a high-level model for how an organization should embed risk management into all its activities. This plan does not contain much detail on specific functional areas and activities of the organization. The plan may contain[21]

- A statement of the organization's uncertainty management policy
- A description of the external and internal context, arrangements for governance, and the unique operating environment within which the organization operates
- Details of the scope and objectives of the uncertainty management activities in the organization, including organizational criteria for assessing whether opportunities and threats are to be avoided or embraced
- Uncertainty management responsibilities in the organization
- A list of opportunities and threats identified and an analysis of the opportunities and threats—usually in the form of a risk register
- Summaries of the uncertainty responses for significant opportunities and threats

The uncertainty management plan should provide fully defined and fully accepted accountability for opportunities and threats, controls, execution, and uncertainty responses for significant opportunities and threats.

Risk management and uncertainty analysis are certainly complicated as some level. The evaluation of traditional risk assessment methods is prone to focus on threats and thereby not include the opportunities. Every

---

BOX 11.1   ESSENTIAL QUESTIONS REGARDING THE CONDUCT OF AN UNCERTAINTY EVALUATION

How do the effects of uncertainty make it difficult to reliably determine the risk associated with the organization meeting its objectives?

Why does risk assessment need to be focused on the effects of uncertainty when *risk* is defined as the "effects of uncertainty" on meeting the strategic objectives of an organization?

Why is it important to include the elements of the uncertainty analysis in the organization's stakeholder engagement process?

Why is it important to have both the opportunities and the threats included in the uncertainty matrix?

How does the uncertainty evaluation help in creating the content for the uncertainty management plan?

---

organization will use some level of what is presented here to make better decisions regarding its opportunities and threats that were determined when preparing the internal and external context. Please consider this information as a general program to guide any organization to have a reproducible means for prioritizing the opportunities and threats so that the activity to respond to opportunities and threats is within the capability of the organization to support.

Essential questions regarding the conduct of an uncertainty evaluation can be found in Box 11.1.

Endnotes

1. ISO, 2009.
2. ISO, 2009a.
3. ISO, 2009.
4. AS/NZS, 2013.
5. AS/NZS, 2013.
6. ISO, 2013.
7. Knight, 2010.
8. ISO, 2013.
9. Knight, 2010.
10. AS/NZS, 2013.
11. ISO, 2009.
12. ISO, 2009.

13. ISO, 2009.
14. Knight, 2010.
15. ISO, 2009.
16. Hillson, 2001.
17. Hillson, 2001.
18. Hillson, 2001.
19. Hillson, 2001.
20. ISO, 2009.
21. AS/NZS, 2013.

12

Sustainable Processes and Operations

Every organization focuses on improving the processes that constitute its operations to help meet its mission and strategic objectives, whether these processes are followed implicitly or explicitly. Effective processes help the organization manage the effects of uncertainty in its internal and external operating environment. Opportunities help organizations achieve the upside of risk when executed well. To become sustainable over time, an organization must successfully identify and act on the opportunities that can contribute significantly to the upside of risk. The most effective way for an organization to optimize its opportunities is to, first, assess all elements of uncertainty with a focus on the organization's environmental, social, and economic responsibilities. These responsibilities form the foundation for meeting the intent of the sustainability policy. Second, the organization must affirmatively address the significant opportunities through the operational processes that were designed to support the strategic and operational objectives. Only in this way can the organization become more sustainable in the long term.

The delivery of efficient operations requires not only diligence in the management of significant opportunities and threats posed by the effects of uncertainty in the internal and external operating environments, but also the recognition of the interests of the stakeholders through both the engagement process and participation in the uncertainty assessment efforts. Efficient operations and processes are necessary to ensure the organization is capable of achieving its mission and strategic objectives. Efficient processes, effective operations, and an efficacious sustainability strategy can be created and embedded in the way work is accomplished in the organization (i.e., the work instructions and operating controls). These processes and operations can then become a major contributor to the sustained success of the organization.

Process Approach

An organization can achieve consistent and predictable results when its operations are understood and managed as a set of interrelated processes. This approach to managing operations aligns all the processes to meet objectives.

It is called a "process approach." A process is a set of interacting activities that use inputs to deliver a product or service. This process approach helps the organization[1]

- Understand the process and be able to consistently meet customer and stakeholder product and service requirements
- Understand how the processes will add value
- Achieve effective processes and efficient operations
- Improve its processes based on the evaluation of data and information collected in the monitoring and measurement activities

The process approach is a management strategy for its product and service operations.[2] By using this strategy, the organization can develop, implement, and control its operational processes.[3] Here are some ways to establish a process approach for an organization's operations:[4]

- Determine and document product and service requirements.
- Establish a process to design and develop products and services.
- Monitor and control external processes, products, and services.
- Manage and control production and service provision activities.
- Implement arrangements to control product and service release.
- Control nonconforming outputs and document all actions taken.

While this list states what needs to be addressed, the organization can tailor the procedures to determine how they will be addressed.

Working with Processes

There are a number of widely used methods that help organizations understand processes so the process approach can be put to good use. In the smallest organizations where operations are still implicit in nature, these methods can be used qualitatively to understand processes. Other organizations can use these methods to begin creating an explicit process approach when meeting the objectives in the face of uncertainty becomes more urgent. Four such methods are described next.

Turtle Diagram

A turtle diagram provides a useful means for understanding a single process (Figure 12.1). This diagram tracks the inputs and outputs of a single

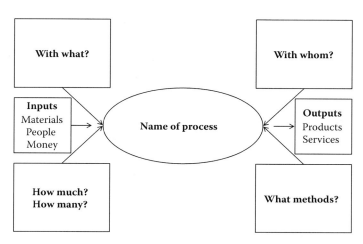

FIGURE 12.1
Turtle diagram for describing a single process—generic concept.

process. It also includes some of the important dependencies that the process relies on:

- What are the required technology, machines, maintenance, materials, infrastructure, and work environment?
- How will it be measured and how often?
- Who are the people and their requisite skills?
- What are the methods and procedures?

Many organizations prepare simple turtle diagrams to use for simple work instructions. Many variations of the turtle diagram in Figure 12.1 can be easily found on the Internet.

Process Mapping

The hierarchical process mapping tool uses multiple layers to depict an entire system of processes responsible for a product or service (Figure 12.2).[5] It is easy to follow the flow since the high-level process has a single digit, along with a description of the work step. The second level has two digits, the third level has three digits, and so forth. The supporting processes can be created using a turtle diagram and then linked to the work steps. Hierarchical process maps are very useful for organizations to improve their processes.[6]

SIPOC Diagram

A suppliers, inputs, process, outputs, and customers (SIPOC) diagram enables the organization to show how the suppliers and customers contribute to the

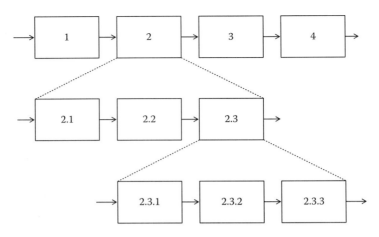

FIGURE 12.2
Multilevel hierarchical process map structure. (Adapted from Pojasek, R.B., *Environmental Quality Management*, 15(2), 79–86, 2005.)

FIGURE 12.3
SIPOC diagram—generic concept.

entire value chain for a process (Figure 12.3). In today's global economy, every organization uses materials from many locations in the world. Sustainability demands that every organization pay attention to the environmental, social, and economic responsibilities of their suppliers. Many organizations are creating "supplier codes of conduct" for suppliers to make certain that all the interests are aligned. These same organizations are also receiving supplier codes of conduct from their customers. There is an expectation that these codes of conduct will be followed, and that each organization will use its sphere of influence to give meaning to sustainability throughout the entire value chain.

Value Chain Diagram

The concept of the value chain[7] conveys the message that processes and activities are universal to all organizations, whether they are a business or have some other purpose. An organizational value chain diagram depicts the organizational processes at the bottom of the diagram and the functional areas that can drive the performance of the processes at the top (Figure 12.4). It takes the combination of and interactions between processes and "people

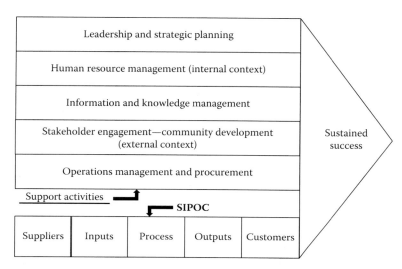

FIGURE 12.4
Value chain diagram—significant change. (Adapted from Porter, M.E., *Competitive Advantage: Creating and Sustaining Superior Performance*, Free Press, New York, 1985.)

processes" to create value and continually improve the processes so that the organization can meets its objectives.

Managing the Processes

A plan–do–check–act (PDCA) method is used to manage processes and the organization's operations as a whole (Figure 12.5). The PDCA method can be described as follows:

- *Plan*: Establish the system of processes, along with the resources needed, to meet the mission, strategic objectives, and operating goals of the organization as it seeks to meet the customer requirements. Planning must also be sensitive to the interests of the stakeholders and the opportunities and threats that were found to be significant in the uncertainty analysis.
- *Do*: Implement the system of processes as were planned.
- *Check*: Monitor and measure the processes, along with the products and services, against the organization's policies, objectives, and requirements. The results of these checks need to be reported in a manner consistent with the organization's policies.
- *Act*: Take actions to improve process performance, as necessary.

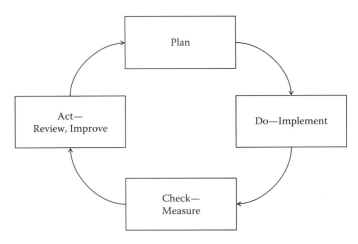

FIGURE 12.5
PDCA process—widely used generic concept.

All systems of management can use this PDCA model for continual improvement of the processes.

According to the principles of risk management,[8] every organization needs to proactively manage its processes so it can have more effective processes and efficient operations. Organizations can use the process approach to establish formality in the more complicated processes, making it easier to determine any interdependencies between processes, look for constraints, and see where resources can be shared. Processes need to be reviewed on a regular basis. When there are problems, suitable actions need to be taken to improve the processes to make them more effective.

As organizations become more complex, the processes need to be managed as a system whereby the networks of processes are created and understood in terms of their sequences and interactions. Consistent operation of such a system is referred to as a "systems approach" to process management.[9] Organizations can develop simple systems for managing the processes; however, these systems must become more rigorous when there are more people and greater demands on the process. This helps people understand what to do and what is an acceptable outcome.

Process Approach and Control

Growth in organizations creates the necessity to accurately determine and plan the processes by which the organization will define the functions necessary for providing the products and services that will meet the needs and expectations of customers, organization members, employees, and other stakeholders on an ongoing basis. It becomes important to plan these processes in light of the organization's strategy, mission, and objectives. This planning enables the control of the processes by which the organization secures the provision of resources, facilitates the delivery of a product or service, monitors the activities, measures the performance of the process, and reviews the operating activities for ways to continually improve them.

Process planning and control within an organization help an organization in

- Watching for changes in the external operating environment
- Forecasting the need for the outputs over the short and medium term
- Determining the interests of the stakeholders
- Understanding the objectives of the processes and how they get translated into the overarching strategic objectives
- Paying attention to the regulatory and legal requirements, as well as the three responsibilities of sustainability
- Managing the uncertainty by managing the opportunities and threats
- Controlling process inputs and outputs, including the use and loss (waste) of resources
- Paying attention to all interactions between processes
- Managing resource productivity and information and knowledge
- Monitoring activities and procedures, along with the people that are competent to perform the activities
- Maintaining records and other documented information required for transparency and accountability
- Conducting proper monitoring, measurement, and analysis of the process outcomes
- Taking corrective action when something does not work as planned
- Continually seeking improvement, innovation, and learning that benefit the outcomes of using the processes

Process planning must be attentive to the determined needs of the organization for developing or acquiring new technologies or developing new products or services, or both. This planning is how organizations add value that enables them to meet their strategic objectives.

Someone must be responsible for a process. Members or employees of an organization must be provided with defined responsibilities and authorities to get the work completed. A "process owner" could be a person or a team of people, depending on the nature of the process and the culture and principles of the organization.

Processes need to be changed from time to time. These changes need to be carefully planned to make certain that they meet the requirements expected of the previous process, while also determining if there are any new requirements or old requirements that will no longer exist after the change. Just as changes in the external operating environment create new opportunities and threats, internal changes also introduce new opportunities and threats. Some internal changes may influence the external context as well. The information gathered by this process of managing change would be kept in the uncertainty (risk) register.

Many organizations restrict their planning activities to operations in a "normal" operating state, but what about emergencies and other uncertain operations or risk events? In an uncertain world, organizations are paying more attention to establishing procedures for determining how to respond to an internal incident, such as a fire, explosion, or accident. Organizations need to pay attention to how to

- Respond to emergency situations and accidents not covered by the characterization of the external context
- Take action to reduce the consequences of many different kinds of emergency situations, appropriate to the magnitude of the incident and the potential for impacts to the external stakeholders and the community at large
- Periodically test the procedures developed for managing these potential situations
- Periodically review and revise the procedures, where necessary, or do this after the occurrence of accidents, emergency situations, or tests of the emergency planning system

It is important to remember that the operations are examined in the internal context, and its associated opportunities and threats should be tracked in the uncertainty analysis.

Resource Management

An organization must consider the internal and external resources that are required to achieve its strategic objectives in an uncertain world. These resources include facilities, equipment, materials, energy, water, knowledge, finances, and people. Resources must be used efficiently so the operations can accomplish "more with less." An organization should identify and assess the uncertainty associated with the securing of certain materials and people when they are needed. This might include scarcity of available resources, as well as mass sickness affecting the members or employees in the organization. The processes need to be optimized in a way that would cover some level of contingency for these and other similar considerations. At the same time, the organization should be vigilant in its search for alternative resources and new technologies. This will lead to a more agile and resilient organization.

Financial resources are reviewed periodically with the organization's financial advisors. This is part of the requirement to maintain sufficient financial resources to support activities required to meet the organization's strategic objectives. Improving the effectiveness of the processes and the efficiency of the operations can improve the financial situation by

- Reducing process, product, and service failures and eliminating materials, waste, and lost employee time
- Eliminating the costs of compensation associated with any guarantees or warranties, legal costs (including mediation and litigation costs), and the costs of lost customers, markets, and reputation

Human resources are very important to the long-term success of an organization. Leaders need to create a shared vision and shared values within the internal context, which will enable people to become fully involved in helping the organization meet its objectives. It is also critical to the success of the organization to provide people with a suitable work environment that encourages personal growth, learning, knowledge transfer, and teamwork. Every organization needs to have a means for managing people through a planned, transparent, ethical, and socially responsible approach. It is important that the members or employees of the organization understand the importance of their contributions and meeting organizational objectives.

Infrastructure must be managed with effective processes so that its operation helps the organizations become more efficient. Such an effort includes

- Maintaining the dependability of the infrastructure needed by the organization
- Ensuring its safety and security
- Providing infrastructure elements needed for products and service provision
- Vigilantly pursuing its efficiency, cost, capacity, and effects on the work environment
- Lowering the impact of the infrastructure and assets on both the work environment and the natural environment
- Identifying and assessing the opportunities and threats associated with the infrastructure to make certain that these uncertainties are

included in the scan of the internal and external operating environment, and in the creation of potential contingency plans for dealing with events that may disrupt the infrastructure

The work environment must be suitable to achieve and maintain the sustained success of the organization, as well as the competitiveness of its products and services. The work environment should encourage productivity, creativity, and a sense of well-being for the people who are working in or visiting the organization's premises. It must also meet applicable statutory and regulatory requirements and address all applicable local standards. A suitable work environment is derived from a combination of human and physical factors that include

- Opportunities for greater involvement in creating processes that help develop the potential of people in the organization
- Safety compliance standards
- Ergonomics
- Dealing with stress and other psychological factors
- Location of the organization
- Proper facilities for members or employees
- Maximization of efficiency and minimization of wastes
- Adequate control of heat, humidity, light, and airflow
- Proper hygiene and cleanliness, while diminishing noise, vibration, and pollution
- Access to nature, preferably physically but at least visually

Attention to the built environment is an important component of an organization's sustainability and should not be handled as a separate "bolt-on" set of initiatives. In a sustainable organization, the built environment should be designed not only for cost efficiencies, but also to provide physical and emotional health benefits and to preserve and protect the environment.

Knowledge, information, and technology need to be treated as essential resources for the organization. The organization needs to identify, obtain, maintain, protect, use, and evaluate its continual need for these resources. It should also share knowledge, information, and technology with its stakeholders, as appropriate. Leaders need to determine the organization's current knowledge and information base and seek to protect that knowledge and information. The need to meet future knowledge and information requirements should also be determined. Reliable and useful data needs to be converted into information for use in decision making. Leaders also need to consider technological options that may enhance the organization's performance in all areas of its operating system.

Natural resources need to be protected to ensure availability of materials in the future to meet the requirements of customers and stakeholders. The organization needs to identify the opportunities and threats associated with the availability and use of energy and natural resources in the short and longer term. Appropriate consideration should be given to the stewardship of its products and services, as well as to the development of designs and processes to address these opportunities and threats. It is important to minimize the environmental, social, and economic impacts over the full life cycle of its products and services, from design, manufacturing, and service delivery to product distribution, use, and end of life.

Efficacious Strategy

The organization's strategy consists of how the organization seeks to meet its objectives in an uncertain world, along with the plans to achieve these objectives. The strategy should be clearly stated in a "looking forward" document. To be efficacious, the strategy must have the power to achieve the desired outcomes. If clear strategic options are created, and vetted through stakeholder engagement and an uncertainty analysis of each of the operational objectives, goals, and action plans, then the result will be an efficacious strategy. The analysis of uncertainty helps prioritize all the identified opportunities and threats found within the operating system, elsewhere in the internal context, and in the external operating environment. Stakeholder engagement and the effective management of uncertainty provide a means for ensuring that the strategy will be efficacious and that strategic decisions will reliably deliver the desired outcomes. Uncertainty analysis and management are critical tasks for creating an efficacious strategy.[10] Uncertainty in the internal and external environment may produce a wide range of different opportunities and threats. The nature of the uncertainty can change almost without notice in a highly dynamic situation. Organizations must be vigilant to capture all the factors from the PESTLE and TECOP analyses and use strengths, weaknesses, opportunities, and threats (SWOT) and other tools to determine and assess the consequences associated with opportunities and threats. Paying close attention to the uncertainty analysis and the development and vetting of the strategy provides the best available information needed by the organization to improve its decision making and its capability to manage effectively.

In an uncertain world, changes in events and other circumstances can reduce the success of the strategy. But if they are included in the uncertainty analysis, then changes in the strategy can be initiated when the sense-making activity detects changes in the context. Then, the organization can determine what controls need to be put in place to optimize the consequences of these changes.[11]

Effective Processes

Effective processes mean that the processes are the correct ones for delivering what is required by the customers and stakeholders. While processes may be efficient, this does not mean that they are correct or the most effective processes that the organization could employ.[12] Developing more effective processes is the way organizations can satisfy customers and stakeholders. The operational system, and its processes and activities, provides the means by which the organization will translate the efficacious strategy into the successful attainment of its objectives. The importance of the operating system in meeting strategic objectives makes it very clear that it is critical to have effective processes.[13] Correct processes are built into the operating system and should never be managed by a set of independent sustainability initiatives.

The uncertainty analysis can help the organization determine the proper response to significant opportunities and threats. Having this information can help identify the necessary operational controls, which will be discussed more in Chapter 13. When implemented, these operational controls must also be evaluated for effectiveness. If the uncertainty management is not comprehensive, the processes will not be effective. The overall intention is to ensure that processes are effective and that operations are efficient at all times. Effective processes provide the means by which operations are changed and strategy is realized.

Efficient Operations

All organizations need to have efficient operations. The best way to ensure the ability of the organization to meet its objectives is to improve the efficiency of the operations. Especially in difficult times, it is important that the operating system continue to be executed as efficiently as possible. As mentioned, delivering more efficient operations can be ensured by developing processes and activities so that they require fewer resources. This creates value for the organization. There is no point in operating efficiently if those operations are based on incorrect processes or separately operated sustainability initiatives.[14] Uncertainty management is also relevant to the continuity of an uninterrupted, efficient operating system. This continuity is a critical component of any sustainability program.

Efficient processes have little to do with "going green." Efficiency and effectiveness are fundamental requirements of an efficacious strategy. Efficiency, effectiveness, and efficacy have been around much longer than "green teams." It is important that sustainability programs facilitate the

> **BOX 12.1   ESSENTIAL QUESTIONS FOR CREATING SUSTAINABLE PROCESSES AND OPERATIONS**
>
> Why is it important to manage the operations of an organization using the ISO process approach?
>
> How does an organization create effective processes throughout its operations?
>
> How does an organization ensure that its operations are efficient?
>
> Why does the sustainability strategy need to be efficacious?

process of maintaining these important elements of a sustainable operating system. The focus on having an award-winning sustainability report or sustainability program will not be the most important contributing factor to enable the organization to meet its objectives in an uncertain world. However, the fact that the organization has an efficacious strategy, effective processes, and efficient operations will be a great story to highlight as a key strategy in the organization's success at meeting its environmental, social, and economic responsibilities, and thereby achieving its sustainability goals.

Essential questions for creating sustainable processes and operations can be found in Box 12.1.

Endnotes

1. ISO, 2008.
2. Praxiom, 2015.
3. Praxiom, 2015a.
4. Praxiom, 2015a.
5. Pojasek, 2005.
6. Pojasek, 2006.
7. Porter, 1985.
8. ISO, 2009.
9. Pojasek, 2005a.
10. Hopkin, 2012.
11. Hopkin, 2012.
12. Hopkin, 2012.
13. Hopkin, 2012.
14. Hopkin, 2012.

13

Organizational Support Operations

People are a critical resource in an organization. Leaders need to create and maintain a shared vision, shared values, and an internal context within which people can become fully involved in meeting the organization's strategic objectives.[1] Organizations need to ensure that the internal context and work environment encourage personal growth, learning, knowledge transfer, and teamwork. The management of people must be performed through a planned, transparent, ethical, and socially responsible manner. People need to understand the importance of their contributions and roles within the organization.[2]

People are involved as operators of the core processes and the supporting processes. Furthermore, people provide organizational support so that everything associated with the operation of the organization will be ensured. However, many of organizational support operations do not get the attention of the sustainability program. Without the range of different support processes, the core processes will not be able to function effectively. The supporting personnel and their value stream processes must be prominently included in an efficacious strategy.

Supporting Human Resources

General

People are a significant resource of an organization, and their full involvement enhances their ability to create value for all stakeholders. Leaders should create and maintain a shared vision, shared values, and an internal environment in which people can become fully involved in achieving the organization's strategic and operational objectives.[3] People management should be performed through a planned, transparent, ethical, and socially responsible approach. The organization should ensure that the people understand the importance of their contribution and roles as a key part of any sustainability program.[4]

Engagement of People

Competent and engaged people are critically essential to the organization's ability to meet its strategic objectives. Engaged people at all levels of an organization can prove beneficial in the following ways[5]:

- Improved understanding of the organization's strategic objectives
- Increased motivation to help achieve the organization's objectives
- Increased involvement in improvement, innovation, and learning activities
- Enhanced personal development and creativity
- Higher levels of satisfaction
- Increased levels of trust and collaboration
- Increased attention to shared values and culture throughout the organization

In order to obtain these benefits, the organization must establish processes that engage people to[6]

- Set the right goals and action plans to enable the attainment of organizational objectives
- Identify constraints to setting and achieving these goals
- Take ownership and responsibility to solve problems
- Assess personal performance against the goals and action plans
- Actively seek opportunities to enhance competence
- Promote teamwork and encourage collaboration among people
- Share information, knowledge, and experience within the organization

Competence of People

To be sure that the organization's members have the necessary competences, the organization should establish and maintain a personnel development plan and associated processes to put the plan into action. The plan should assist in identifying, developing, and improving the competence of people through the following steps[7]:

- Identifying the professional and personal competences the organization could need in the short and long term
- Identifying the competences currently available in the organization and the gaps with what is needed in the future
- Implementing actions to improve or acquire competences to close the gap

- Reviewing and evaluating the effectiveness of actions taken to ensure that the necessary competencies have been acquired
- Maintaining competences over time as needed

The organization needs to determine the knowledge necessary for the operation of its core processes that produce products and services (see Chapter 3). This involves both its current knowledge and its ability to acquire or access the necessary knowledge as the processes and operations are improved and maintained in a compliant condition. Some of this knowledge is derived from internal sources (e.g., learning from successes and failures), and some is derived from external sources (e.g., standards, schooling, conferences, and engagement with customers and other stakeholders). Knowledge is a key element that supports decision making at all levels of the organization.[8]

Next, the organization must determine the necessary competence needed for working within the system of management and affecting organizational performance and effectiveness. It is important to ensure that people within the organization are competent on the basis of their education, training, and experience. When necessary, actions should be taken so that people can acquire the necessary competence and so that leaders can evaluate the effectiveness of these actions.

People working in an organization need to be aware of the policies, objectives, and goals that relate to their work, and how they contribute to the effectiveness of the system of management. This awareness helps people understand the benefits of using their knowledge and experience to help the organization meet its strategic objectives. Awareness also helps people understand the negative implications of not conforming to the system of management.[9]

People learn about the progress toward meeting the strategic objectives through effective, two-way dialogue and communication. Leaders need to provide an appropriate emphasis on the process of engagement. Providing only mechanistic internal and external communication is not as effective or enduring.

Internal Value Chain Support

The internal value chain has a number of functions that support the operations with processes associated with the activities of people. It is important to have these people-focused support processes be as effective as the processes used in operations (i.e., which are directly associated with products and services). These support activities include

- Leadership
- Strategic planning
- Employee engagement

- Stakeholder engagement
- Information and knowledge management
- Operations management

These support activities are found in many performance-focused programs (see Chapter 16). Each of the categories above has criteria that represent some level of "best practice." Organizations can self-assess against the criteria for the purpose of comparing their support processes against the best practices of organizations in each category.[10]

Communication

The organization should determine the need for internal and external communication. It is helpful to view communication as the "messages" that inform the ongoing engagement process. Engagement is a form of dialogue not restricted to communication. Sustainability must be an integral part of this communication and not conducted separately. A communication program should consider[11]

- What it will communicate
- When it will communicate
- With whom it will communicate
- How it will communicate

Internally, the organization should adopt appropriate methods of communication to ensure that the sustainability message is heard and understood by all employees on an ongoing basis. Much of the communication provides information on the core and supporting operations. The communication should clearly set out the organization's expectation of employees. External communication should provide messages to the stakeholder engagement process. The entire range of stakeholders should receive all relevant external communications.

Communication should be specifically designed to enhance awareness of all the organization's stakeholders. This awareness should include

- The sustainability policy and other policies of interest to each party
- Relevant strategic, operational, and tactical organizational objectives
- Their role and contribution to effective processes, efficient processes, and efficacious strategy
- The implications for not conforming to the system of management requirements

Supporting Information Resources

Information Support

The organization should establish and maintain processes to gather reliable and useful data and convert that data into the information needed for decision making. Information support includes the processes needed for the storage, security, protection, communication, and distribution of data and information to all engaged stakeholders.[12] An organization's information and communication systems need to be robust and accessible to people making decisions at all levels of the organization. It is important for the leaders to do what is necessary to ensure the integrity, confidentiality, and availability of information that relates to the organization's performance, process improvements, and progress toward the achievement of sustainability.[13]

Compliance Information

Every organization needs to determine and provide the resources necessary for the establishment, development, implementation, evaluation, maintenance, and continual improvement of all environmental, social, and economic compliance information.[14] The information itself and the manner in which it is managed are related to the size, complexity, structure, and core and supporting processes. Leaders are expected to make available the necessary compliance-related resources and deploy them effectively to ensure that compliance management meets the organization's strategic and operational objectives, and that compliance is achieved.[15] These resources include financial and human resources, as well as access to expert and skilled people who can consult with the organization based on their knowledge, professional development, and information technology. These experts may have legal, engineering, management system, process improvement, or experience credentials and related information technology.

Monitoring and Measuring Resources

The organization should identify the internal and external resources that are needed for the achievement of the organization's strategic and operational objectives in the short and long term. To ensure that these resources (e.g., equipment, facilities, materials, energy, knowledge, finances, and people) are used effectively and efficiently, the leaders need to have processes in place to provide, allocate, monitor, evaluate, optimize, maintain, and protect those resources.[16] To ensure the availability of the resources for future activities, the organization needs to identify and assess the threats of potential scarcity and continually monitor the current use of economic, material, and people resources to find opportunities for the optimization of their use. This should be done in parallel with research for new resources, optimized processes, and new technologies.[17] The organization needs to monitor and measure the availability and suitability of the identified resources, including outsourced resources, and take action as necessary. The results of the monitoring should also be used as inputs to the process of maintaining the organization's efficacious strategy.

Organizations use monitoring and measuring to ensure that valid and reliable processes are used for their products and services. In some cases, these resources include measuring equipment[18] that is suitable for the activities involved and maintained to ensure fitness for its purpose.

In some processes, there is a requirement for measurement traceability, especially when it is considered to be an essential part of providing confidence in the validity of the measurements.[19] Measurement standards are used for this purpose. When no international or national measurement standard exists, the basis used for calibration or verification is retained as documented information. All measuring equipment and standards are protected from adjustments, damage, or deterioration that would invalidate the instrument calibration and subsequent measurement results.[20] More information on monitoring and measurement is provided in Chapter 16.

Financial and Risk Management Information

Leaders should determine the organization's financial needs and establish the necessary financial resources for current and future operations.[21] The organization needs to establish and maintain processes for monitoring, controlling, and reporting the effective allocation and efficient usage of financial resources related to the organization's strategic objectives. Reporting of the financial information can also provide the transparency required for determining ineffective or inefficient activities, while receiving feedback on the financial justification for initiating suitable improvement actions. Financial reporting should be used in all management reviews and in the evaluation of the effectiveness of the sustainability efforts. Improving the effectiveness and efficiency of the organization's operations and supporting core operations can positively influence the financial results.

There are areas where supporting processes and the mechanical and information assets need to be carefully monitored by the financial and risk management (e.g., insurance provision) functions of the organization.[22] A key element in this sharing of information involves life cycle costing analysis and the costing levels of service options. This information is required for benefit analysis that will be used to evaluate the entire set of operations—core operations and supporting operations.

Many insurance carriers seek information on specific equipment or practices that could lead to financial exposures. The concerns arise with many of the issues mentioned in this chapter. By managing these concerns, the processes should be more effective and the amount of money spent on insurance should decrease over time.

Knowledge Management Information

An organization needs to identify the knowledge necessary to improve the effectiveness of the processes and the efficiency of the core and supporting operations. The knowledge management processes should address how the organization identifies, obtains, maintains, protects, uses, and evaluates the need for these knowledge resources. Leaders need to assess how the organization's current knowledge base is identified and protected. Leaders should also consider how to obtain the knowledge required to meet the present and future needs of the organization from both internal and external resources.[23]

There are many ways to identify, maintain, and protect knowledge[24]:

- Learning from failures, near-miss incidents, and successes
- Capturing the knowledge and experience of the people in the organization
- Gathering knowledge from customers, stakeholders, suppliers, and partners
- Capturing undocumented knowledge (tacit and explicit) existing in the organization
- Capturing important information from documented data and records

This information is vital to their ability to offer conformant products and services. Knowledge is also required for the improvement of decision making at all levels in the organization (see Chapter 3).

Documented Information

Documented information is created by the organization based on what is necessary for ensuring the effectiveness of the system of management. Obviously, this will differ substantially with the size of the organization and its activities, processes, products, and service, as well as with the competence of the people within the organization.[25]

Creating and updating documented information needs to be performed in a consistent and reliable manner. There is a need for the identification and description (e.g., title, date, author, or reference number), a useful format (e.g., wording, software version, and graphics), and the periodic review and approval of the documentation for suitability and adequacy.[26]

Finally, documented information needs to be carefully controlled so that it is available where and when it is needed and protected (e.g., from improper use, loss of confidentiality, or loss of integrity).[27] For the proper control of documented information, the organization needs to address its distribution, access, retrieval, and use. There must be a provision for proper storage and preservation of the documents. It is important to maintain both version control and legibility of the documents, along with plans for the retention or destruction of the documents when it is deemed they have served their useful or legal purpose. Often, there are some documents that come from outside of the organization (e.g., standards, customer methods requirements, and service manuals). The degree to which these are managed needs to be determined and assessed by leadership and then included as part of the control function.[28]

Information on Physical Assets

Infrastructure

The organization needs to provide the infrastructure necessary to support the operations discussed in Chapter 12, by planning, providing, and managing its infrastructure in an effective and efficient manner to help meet its operational objectives. The infrastructure supports the organization's ability to be able to provide products and services. Examples of infrastructure include the following:

- Buildings; storage areas; tank farms; roadways; electricity; water; compressed air; fire suppression; compressed gases; heating, ventilating, and air-conditioning; and filters
- Equipment, including surveillance systems, computers, microprocessors, and software
- Transportation, including loading docks, forklifts, containers, trucks, conveyors, and elevators
- Information and communication technology, including speaker systems and alarm systems

Infrastructure is maintained as part of maintenance control in the assets management system.

The process of managing the organization's infrastructure should provide consideration of the following[29]:

- Dependability of the infrastructure
- Safety and security
- Efficiency, cost, capacity, and work environment
- Impact of the infrastructure on the work environment

Organizations need to identify and assess the risks associated with the infrastructure and take appropriate action to manage uncertainty, including the establishment of adequate preparedness and response plans.[30]

Technology

Leaders should consider technology options to enhance the organization's performance in areas such as operations, supporting operations, marketing, benchmarking, customer and stakeholder interaction, supplier relations, and all outsourced processes.[31] The organization should establish effective processes for the assessment of[32]

- The current levels of technology usage inside and outside of the organization, including emerging trends
- Economic costs and benefits
- The evaluation of opportunities and threats related to changes in technology
- The competitive environment
- Its speed and ability to react to customer and stakeholder interests

Working Environment

The organization needs to determine, provide, and maintain the work environment necessary to support the quality of the products and services. In addition, the work environment should encourage productivity, creativity, and well-being for the people who are working on or visiting the organization's premises. Leaders should seek to provide a suitable work environment to achieve and maintain the sustained success of the organization and to support the competitiveness of its products and services. There are a number of human and physical factors that need to be addressed[33]:

- Optimal social conditions determined by the social well-being responsibility of the sustainability program, including human rights, nondiscrimination, proper discipline actions, safety rules, and guidance for the use of protective equipment
- Facilities for the people in the organization
- Physiological factors, such as stress reduction and fatigue recognition
- Ergonomic issues, such as repetitive motions
- Physical elements, such as heat, cold, humidity, light, airflow, personal hygiene, noise, vibration, and pollution
- Maximization of efficiency and minimization of waste

Many of these factors and situations are regulated by environmental, occupational health, and safety requirements. The factors in the workplace differ substantially depending on the products and services provided and the internal and external context of the organization.

> **BOX 13.1   ESSENTIAL QUESTIONS ON**
> **SUPPORT PROCESSES AND PEOPLE**
>
> How do people support the various processes responsible for producing the organization's products and services?
>
> Why do the value chain support personnel need to pay attention to the engagement and competence of people responsible for processes that deliver products and services?
>
> How do the value chain support personnel help manage the supporting information resources?
>
> How do the value stream support personnel help manage the physical assets and the working environment?

Availability of Natural Resources

The availability of natural resources is a key concern in a sustainability program. The organization needs to consider the opportunities and threats related to the availability and use of energy, water, and other natural resources in the short and long term.[34]

An organization needs to consider the integration of environmental protection aspects into product and service design, as well as the development of its processes to address opportunities and threats. It is also important to seek to minimize environmental impacts over the full life cycle of its products, services, and infrastructure, from design, through manufacturing or service delivery, to product distribution, use, disposal, or other end-of-life options.[35]

Essential questions for developing support processes and people can be found in Box 13.1.

Endnotes

1. ISO, 2009b.
2. ISO, 2009b.
3. ISO, 2009b.
4. ISO, 2009b.
5. ISO, 2014.
6. ISO, 2009b.
7. ISO, 2009b.
8. ISO, 2009b.

9. ISO, 2009b.
10. Pojasek and Hollist, 2011.
11. ISO, 2015.
12. ISO, 2009b.
13. ISO, 2009b.
14. ISO, 2014b.
15. ISO, 2014b.
16. ISO, 2009b.
17. ISO, 2009b.
18. ISO, 2015.
19. ISO, 2015.
20. ISO, 2015.
21. ISO, 2009b.
22. ISO, 2014c.
23. ISO, 2009b.
24. ISO, 2009b.
25. ISO, 2015.
26. ISO, 2009b.
27. ISO, 2009b.
28. ISO, 2009b.
29. ISO, 2009b.
30. ISO, 2009.
31. ISO, 2009b.
32. ISO, 2009b.
33. ISO, 2009b.
34. ISO, 2009b.
35. ISO, 2009b.

14

Organizational Sustainability

To be successful, sustainability must be developed from within the organization, by building on what is already in place. Sustainability should create relationships between processes and operations having regard for the interests of the stakeholders. The organization incorporates sustainability within its system of management to encourage resource productivity, decision making based on factual evidence, and a focus on the customers for its products and services, as well as on the interests of the stakeholders and the community at large. Organizations should seek an appropriate level of sustained success in line with the complexity of the decisions that need to be made. An organization also needs to have a *current* and *comprehensive* understanding of its uncertainty (i.e., opportunities and threats) and the risks associated with meeting its strategic objectives.[1]

Attributes of Organizational Sustainability

There are eight attributes that organizations can use to evaluate the extent to which sustainability is effective and whether it can be improved. Organizations should aim at an appropriate level of performance of their risk management and sustainability programs that is in line with the complexity of the decisions that need to be made.[2] This list of attributes represents organizational sustainability at a high level.

Ethical Behavior

An organization should behave ethically. This behavior stressed values of honesty, equity, and integrity. These values imply a concern for people, animals, plants, and the environment, as well as a commitment to address the impact of its activities and decisions in light of its stakeholder's interests.[3] This attribute can be tested by examining how the organization defines and communicates the standards of ethical behavior expected from its governance structure, personnel, suppliers, contractors, and owners and leaders. Usually, this information is found in the organization's "code of conduct," or it may be found in the customer's "supplier code of conduct," which is often made part of the purchase order for products and services.

Respect for the Rule of Law

An organization must demonstrate its respect for the rule of law. It is generally implicit in the rule of law that laws and regulations are written, publicly disclosed, and fairly enforced according to established procedures. In the context of sustainability, the organization should take steps to be aware of applicable laws and regulations, and to inform those within the organization of their obligation to observe and implement those measures.[4] This is tested by determining how the organization keeps itself informed of all legal and other obligations and ensures that its relationships and activities comply with the intended and applicable legal framework. The organization should also conduct a periodic evaluation of its compliance with applicable laws and regulations, including contractual adherence to codes of conduct.

Respect for Stakeholder Interests

An organization must respect, consider, and respond to the interests of its stakeholders. Besides the internal stakeholders, other individuals or groups may also have rights, claims, or specific interests that need to be taken into account. Sustainability also demands that the organization consider the interests of stakeholders who may be affected by a decision or activity, even if they have no formal role in the stakeholder engagement efforts or are unaware of these interests.[5] This is tested by the ability of stakeholders to contact, engage with, and influence the organization. There should also be documented information on the role stakeholders play in the conduct of the organization's uncertainty analysis.

Full Accountability for Opportunities and Threats

Uncertainty analysis includes comprehensive, fully defined, and accepted accountability for testing opportunities and threats for significance and taking the proper response. All members or employees of an organization need to be fully aware of the opportunities and threats, the controls and engagement tasks for which they are accountable. The definition of risk management and sustainability roles, accountabilities, and responsibilities should be part of all the organizations work instructions and operating controls.[6] Many organizations include the performance of uncertainty assessment and sustainability accountabilities in performance reviews.

Engagement with Stakeholders

Risk management and sustainability require continual, interactive communications with external and internal stakeholders, including comprehensive and frequent reporting of uncertainty assessments (i.e., opportunities and threats), and the application of risk management and sustainability to an appropriate degree. Communication is seen as a two-way process, so that properly informed decisions can be made about the level of uncertainty and the need for opportunity and threat response in line with properly established and comprehensive uncertainty criteria.[7] This can be tested by looking at the documented information on the stakeholder engagement process.

Embedding Sustainability in the Organization's Governance

Risk management and sustainability are considered to be central to an organization's system of management, such that risks of meeting the organization's strategic objectives are improved by managing the effects of uncertainty (i.e., opportunities and threats). The governance structure and process are based on the management of risk, uncertainty, and sustainability. Effective management of the opportunities and threats is regarded by leaders as essential for the achievement of the organization's strategic objectives[8] in an uncertain world. This is indicated by leaders' language and the written materials in the organization using the term *uncertainty* in connection with risks to the strategic objectives.

Including Sustainability in Decision Making

All decision making within the organization, whatever its level of importance or significance, involves the explicit consideration of the effects of uncertainty (i.e., opportunities and threats) and the management of risk of meeting its strategic objectives. This can be tested to determine if performance evaluation criteria include sustainability in the decision making of individuals in the organization. There should be documented information to determine whether decision making is expected to include explicit consideration of the organization's strategic objectives, the effects of uncertainty, organizational knowledge, risk management, and sustainability.

Continual Improvement and Learning

An emphasis is placed on continual improvement in risk management and sustainability through the setting of goals and action plans against which the organization's members assess whether the strategic and operational objectives are being met. The organization's performance can be documented and shared with the stakeholders[9] during the engagement process. Sustainability often has a regularly scheduled management review meeting as a component of its system of management. The learning is expressed as the "act" component of the plan–do–check–act cycle.

Sustained Success of an Organization

To achieve sustained success, leaders should adopt a risk management and sustainability approach. The organization's system of management should be based on the sustainability principles found in Chapter 2. These principles describe concepts that are the foundation of an effective risk management and sustainability program.

The organization achieves sustained success by consistently addressing the interests of its stakeholders, in a balanced way, over the long term.[10] An organization's internal and external operating environment is ever changing and uncertain. To achieve sustained success, the organization's leaders should[11]

- Maintain a long-term planning perspective
- Constantly monitor and regularly analyze the organization's operating environment
- Identify all stakeholders and assess their potential impacts on the organization's performance
- Engage stakeholders and keep them involved in the review of the organization's management of its significant opportunities and threats
- Establish mutually beneficial relationships with members of the organization's value chain
- Conduct uncertainty assessments as a means of identifying and managing the significant opportunities and threats
- Anticipate future resource needs and the overall resource productivity of the organization
- Establish processes appropriate to achieving the organization's strategic objectives
- Establish and maintain processes for continual improvement, innovation, and learning

These processes for sustained success are applicable to any organization, regardless of size, type, and activity. Attention should also be paid to a number of important risk management and sustainability approaches to different elements found in organizations.

Practices for Embedding Sustainability in an Organization

Risk management and sustainability are critical components of an organization's sustained success. These practices should be embedded in the decision-making process by including them in the knowledge, policies, organizational culture, strategies, and operations. The organization needs to build internal competency for sustainability, undertake engagement on sustainability with all stakeholders, and regularly review its actions and practices using the attributes of organizational sustainability.

Importance of Organization's Context

To begin the embedding process, it is important to make sure that the mission statement has been converted to a clear set of strategic objectives, and that the cascading of the objectives from the top down is in place. To provide the feedback, the goals and action plans must also be in place. These practices are described in Chapter 7. Once this happens, the organization will have the means for the sustainability program to be embedded in the process for meeting the strategic objectives, rather than having a separate set of "sustainability goals" with supporting initiatives that are not formally coordinated with the organization's objectives.

The scanning of the organization's internal and external context helps identify its opportunities and threats. Sustainability practitioners need to be involved in this search for opportunities to make sure that they are properly identified. This is a key linkage for sustainability to make as it starts the embedding process. The scanning of the external operating environment provides the best view of the social, environmental, and economic requirements of the organization. Examining the legal and other requirements would be a part of this effort. Some organizations have contractual commitments to follow their customers' supplier codes of conduct. There may be other contractual requirements to customers, as well as a parent operation.

Scanning of the internal and external operating environments also helps identify the stakeholders. These individuals and organizations are associated with the opportunities and threats identified in this activity. Sustainability is usually outwardly focused. Now that the opportunities and threats are associated with the uncertainty, there is an internal need for sustainability to help lower the amount of uncertainty so that the risk of meeting the strategic objectives is improved. Sustainability can make sure that the stakeholders are engaged and that there is a proper sense-making effort in place to determine changes in the internal and external operating environments that would dictate when a new scan is needed.

This scanning activity can be expanded within the organization's value chain. Suppliers, contractors, customers, and business partners need to be engaged through sustainability to manage uncertainty associated with these relationships. All the members of the value chain need to conduct their own scans of the internal and external operating environments. Uncertainty can come from any point in the value chain. Part of the relationship is to provide some support to value chain partners to help lower the uncertainty of the entire value chain to promote the "upside of risk," where the opportunities are offsetting the threats to manage risk within a tolerable or even beneficial level.

It is also important for an organization to be aware of the level of commitment of the leaders to use sustainability, along with the other organizational risk management methods, to make sure the organization is always in a comfortable position for meeting its strategic objectives. Where there is an active focus on meeting organizational objectives, it is easy for the leaders to recognize the roles played by the different functional units in keeping the level of risk in line with expectations. When sustainability is seen as an active participant in managing uncertainty and risk, it is much easier to obtain the approval of leadership. The contribution of sustainability to meeting strategic objectives becomes clear when it is embedded in these processes.

Understanding Organizational Sustainability

A key role for sustainability in an organization is participation in the due diligence activities. This is a process to identify the actual and potential positive and negative social, environmental, and economic impacts of an organization's decisions and activities. This happens with the active management of the internal and external context. It is no longer the case of just looking to avoid and mitigate the negative effects (threats), but it is more about using the positive effects (opportunities) to create a positive upside of risk. This due diligence also involves *influencing* the risk and uncertainty analysis conducted by other organizations in the value chain. With the speed of news in this digital media world, a lot of negative publicity can happen all of a sudden when a supplier has encountered problems. Increased communication and cooperation within the value chain is now seen as critical to the long-term success of any organization, of any size or type.[12]

When the organization conducts its uncertainty analysis, it identifies the opportunities and threats that are most *significant* to the organization. This is a formal process that can be either qualitative or quantitative. An uncertainty (i.e., risk) map is used to identify the organization's opportunities and threats altogether. Stakeholders need to be involved in this process. If the engagement process is working well, the stakeholders will be a valuable source of information for the organization. They can help alert the organization to changes in the external operating environment and the need for an update of the uncertainty analysis. It is possible that they can help in the offsetting of the threats with the opportunities that have been identified.

In the areas of sustainability and corporate social responsibility, this is often referred to as a "materiality" determination. Rarely do these determinations use uncertainty analysis and tie the results into the way risk is managed within the organization and its value chain. So, do not confuse the formal uncertainty analysis with a materiality determination.

One area that sustainability programs find to be difficult involves the determination that the organization is in compliance with local, state, and federal laws and regulations that govern its operations. A careful review of the opportunities and threats often indicates that there are some nuisance areas that involve odors, vibrations, light, noise, and traffic, among other things. While these might meet the legal tests, the situations can be very irritating to the stakeholders. Likewise, a product sold in a retail store might be made in a country that has serious human rights issues or environmental regulations that are not enforced. These kinds of interests should be found in the scanning activities; however, there is a sense of complacency associated with meeting the requirements. This is just one more reason to have a strong stakeholder engagement program. Stakeholders can help seek opportunities, as well as threats.

Embedding Sustainability

Embedding sustainability into every aspect of an organization involves attention to the items covered in the elements presented in this chapter. This is much more effective than trying to add sustainability to a long list of expectations for members or employees. The approach to sustainability should make it a part of what everyone does every day and use its influence throughout the internal and external context of the organization.

It becomes easier to engage the leaders in sustainability when it is linked to risk management. As described in Chapter 5, risk and uncertainty management are almost always embedded in the organizations since meeting the strategic objectives is so crucially important to any organization. Building the competency for using sustainability can become a part of the organization's drive for effective and efficient processes. Creating a culture of sustainability within the organization is easier when everyone sees it as a way to meet their goals so that their department can meet its objectives. Meeting strategic objectives benefits everyone associated with an organization. Risk management and sustainability contribute to the ability of the organization to have an efficacious strategy.

As risk management and sustainability become embedded in the organization, there may be a need for changes in decision-making processes at all levels of the organization. There may also be some changes in governance that will recognize the positive consequences of embedding risk management and sustainability into the processes used by the organization and its partners in the value chain.

There are some that feel that sustainability needs to be "integrated" into the organization, and that this will take a lot of time and effort. By embedding sustainability as described above, it is embraced more effectively and quickly. However, this is often threatening to the sustainability manager.

Some compare sustainability to quality. When an organization is recognized for the quality of its products and services, it is not a time to cut back on quality. With a global economy and the need for continual improvement, the lead in quality can be lost suddenly. This is much less likely to happen when all the disciplines are embedded into the processes and operations to help the organization function effectively and efficiently and to maintain its social license to operate.

Engagement, Not Simple Communication

Communication must be effective two-way dialogue over a period of time in order to become engagement. Many organizations continue to use traditional means of communication to develop awareness about the role of risk management and sustainability in helping an organization meet its strategic objectives. Frequent conversations on how the organization can improve the use of sustainability in its processes and operations need to be interactive and acknowledged. Digital media is helping make this more effective. The dialogue must be transparent and thoughtful. It is important to have a focus on helping the organization meet its responsibly set objectives over the long term. If objectives need to be changed, there needs to be a dialogue over the pros and cons of such a change and the linkage of the change to risk management and sustainability.

Communication should be part of the processes and used in the operations. It exists to inform people and transmit important messages. Engagement takes place both within the organization and with the external stakeholders. Often, it may be an external stakeholder that starts the engagement with a question, complaint, or suggestion. It could be started by the organization to explain the need to make some changes and by inviting others to participate in the decision-making process.

The people within the organization need to become familiar with the different methods and media that may be used for engagement. Thoughtful answers come from discussions with others before responding. This allows other people to join in the dialogue.

A special form of engagement is with the media. Many organizations that deal with the media on a regular basis get to know someone at the media outlet and invest time to get to know him or her better. The building of trust will help enable some engagement to commence. The focus should always be on how the organization is able to meet its strategic objectives while providing shared value with the external stakeholders. It makes a lot of sense to want a vibrant community and make an investment to see that happens when you know that employees are more likely to remain with the organization when the community is very strong and healthy. In these ways, there is a difference between handling sustainability as a separate program and having sustainability embedded in the organization and made part of what every employee does every day.

Reviewing and Improving Organizational Sustainability

Effective performance on risk management and sustainability depends on commitment, participation, engagement, evaluation, and review of the contributions to effective processes, efficient operations, and efficacious strategy. Monitoring and measurement related to sustainability will come from the people that are engaged in using them to do their work every day. There will also be a means for collecting the information on how the significant opportunities and threats were managed to enable the organization to be on the upside of risk.

Living in an uncertain world makes it imperative to monitor the external operating environment and the entire supply chain. These numbers are not created for a report. Instead, the measurements are part of the continual improvement, innovation, and learning in an organization.

Contrary to popular thought, one does not just pick processes and operations and measure them. There is a formal process of monitoring and measurement that needs to take place. It will produce information for operating the organization and engaging with the external stakeholders. The information will be transparent, and the organization will be accountable for improving so that it can meet its strategic objectives.

The third section of this book is focused on the monitoring and measurement that are necessary to drive the organization's ability to meet its strategic objectives over the long term. Monitoring and measurement need to be understood by everyone in the organization, and each person must use facts to meet their goals in order for the organization to meet its strategic objectives.

Essential questions for implementing organizational sustainability can be found in Box 14.1.

BOX 14.1   ESSENTIAL QUESTIONS ON ORGANIZATIONAL SUSTAINABILITY

Why is it so important to establish multiple connections between the various structural elements of a sustainability program as discussed in this chapter?

How do the attributes of organizational sustainability help an organization make these connections between the structural elements?

How does this chapter help make it clear that structural elements are needed to define a sustainability program and embed it in the way the organization operates every day?

Why is it important to have a monitoring and measurement program to determine how effectively this integrated structure is working?

Endnotes

1. AS/NZS, 2013.
2. AS/NZS, 2013.
3. ISO, 2010.
4. ISO, 2010.
5. ISO, 2010.
6. AS/NZS, 2013.
7. AS/NZS, 2013.
8. AS/NZS, 2013.
9. AS/NZS, 2013.
10. ISO, 2009b.
11. ISO, 2009b.
12. ISO, 2010.

Section III

Monitoring, Measuring, and Improving Organizational Sustainability

Section II of this book explained the "plan" and "do" elements of the plan–do–check–act (PDCA) cycle for implementing and maintaining a sustainability program in an organization. Chapters 15 through 19 address the "check" and "act" elements of this widely used PDCA method.

Sustainability has always had a focus on measurement so that different organizations can demonstrate their transparency and accountability for implementing essential features of the program and seek to continually improve that program. The information in each of these chapters is linked to best practices that can be traced using the endnotes and references. As in Section II, the focus is on "what" it is that needs to be accomplished. The reader can determine how this should be accomplished for a particular organization. It is important that the same level of rigor be imposed in this effort that was outlined in Section II. Section I provided some foundation work that will support this effort to monitor and measure sustainability so that the organization can continuously improve the embedded sustainability program.

Appendix III contains the "check" and "act" elements for a virtual hotel. The reader can practice using the information in this book to improve the initial effort found in that case. Attention should also be paid to determining if the essential questions at the end of each chapter are addressed in the revised case. Completing this work will also lead to the need for further improvements in the Section II contents. Such is the world of continual improvement of an embedded program.

Chapter 20 shows how to adapt the practice of resilience into the same PDCA framework. Many people feel that resilience is the next big wave after sustainability.

15

Scoping the Monitoring and Measurement Process

Measurement is a quantitatively expressed reduction of uncertainty based on one or more observations.[1] The observations are made as part of the monitoring process. If something can be observed in any way, it lends itself to some type of measurement method. And those things most likely to be deemed immeasurable are often addressed by relatively simple measurement methods.[2]

What needs to be measured? What methods should be used for monitoring, measurement, analysis, and evaluation to ensure valid results? The organization needs to evaluate the performance and effectiveness of its sustainability program. Measurements are relevant and critical to major organizational decisions and help the organization track its progress toward meeting it strategic objectives. Yet, we do not seem to find an obvious and practical means to conduct the relevant monitoring and measurement tasks in a sustainability program. Sustainability practitioners are focused on gathering available data and reporting the information to interested stakeholders in the external operating environment. Understanding monitoring and measurement is critical to the success of sustainability and a key component in helping an organization meet its strategic objectives. Measurement also supports the organization's pledge to continual improvement.

Why Does an Organization Measure?

Measures are needed to guide the organization to the realization of its strategic objectives and mission statement. At each level of the organization (i.e., strategic, operational, and tactical), a family of measures is needed to create a complete picture of that level's strengths and weaknesses so that appropriate decisions can be made with the correct follow-up actions taken.[3] Typically, the leaders are focused on the financial viability of the organization. People at the tactical level are focused on their goals and action plans (see Chapter 7). Their measures are watched every day and are measured in physical amounts: units, pounds, inches, minutes, and components. Meeting the action plans is the driving force. The middle of the organization uses screening or diagnosis measures. Is the data from the tactical level demonstrating that the tactical and operational objectives are being met? Priorities on the measures are always set according to the interests of the stakeholders.

Some of the key decisions for using measures include[4]

- Any action, process, or operation can be measured.
- Consistent attention must be paid to consistency of measures to determine trends.
- What is done with the measures is as important as what the measures are.
- Stakeholder interests should be closely related to the strategic objectives.
- Measurements must look at both lead and lag indicators. Lead indicators are the performance drivers that communicate how results (lag indicators) are to be achieved.
- There is more to organizational success than financial success.

All these decisions help an organization manage by fact. The "right" measure is the one that tells an organization's leader whether the stakeholder interests that are critical to the long-term success of the organization are on target.

Deciding What to Measure

Organizations need to have effective systems and processes for determining what data and information should be collected, and how they are handled, stored, analyzed, and interpreted.[5] This information is used to increase the organization's understanding of the internal and external operating environment within which it operates. Data and information need to be continually reviewed to ensure they remain current, meaningful, and effective.

The organization should establish criteria and processes for determining what data should be collected and for not collecting data that is not useful for these purposes. Processes must have links between data gathering and the purpose for gathering the data. Sources of data need to be reviewed on a regular basis to monitor their effectiveness and continual improvement in meeting changing organizational requirements. Based on the scanning of the external operating environment, a wide range of stakeholders are involved in determining what data should be gathered.[6]

Organizations provide for the analysis and interpretation of the data in order to learn and inform its decision making in both the short and longer term. The analysis of the data helps the organization understand the nature and impact of variation on processes, products, services, and measurement systems. Data is also used as the basis for maintaining awareness among the internal stakeholders and creating incentives for innovation.[7]

An organization needs to establish systems to ensure that data is shared among those that need to use it to improve performance, and that the data is generally accessible. Data collection needs and methods must be reviewed on a regular basis. It is also important to review the sampling methods and sampling systems.[8] These responsibilities help ensure that the data is valid, reliable, relevant, timely, secure, and sound.[9] The measures or indicators selected should best represent the factors that lead to improved customer, operational, financial, and societal (i.e., external stakeholders) performance.[10]

Sustainability managers need to figure out what sustainability is supposed to do when it is embedded within an organization. What does sustainability mean to processes and operations, and why does it matter? Sustainability is a vague concept to many people that needs to be associated with what is actually expected to be observed. Once everyone is familiar with the answers to these questions, the situation starts to look more measurable.[11] The decision-making for sustainability would progress as follows[12]:

- If sustainability matters, it is likely to be detectable and observable.
- If it is detectable, it can be detected as an amount or range of amounts.
- If it can be detected as a range of possible amounts, it can be measured.

It is also important to state *why* it is important to measure sustainability in order to understand *what* is really being measured. Understanding the *purpose* of the measurement is often the key to defining what the measurement is really supposed to be. Some things seem to be "intangible" only because people have not defined what they are talking about.[13]

Measurement Methods

Measurement involves using different types of sampling and experimental controls. This involves taking small random samples of some activity. By measuring the situation that is of interest, it is possible to learn what to measure. It is important to get past the point where decision makers seek to avoid making observations by offering a variety of objections to the measurements. To make some productive progress, consider four useful measurement assumptions[14]:

1. The problem is not as unique as previously thought.
2. There is more data than originally thought.

3. Less data is needed than originally thought.

4. An adequate amount of new data is more accessible than previously thought.

Think of measurement as iterative. Start measuring what sustainability is contributing to the organization. You can always adjust the method based on the initial findings. Making a few more observations can help you learn more about what you are trying to measure and for what reason.

Characteristics of a Measurement System

Even though there is much disagreement, the prevailing thought on the maximum number of metrics for an organization is around 20. Think about it. How many things can be monitored and controlled on a regular basis? This is why key performance indicators (KPIs) were invented. Some larger organizations take thousands of metrics and roll them up to a small number of KPIs. Organizations are always advised to see their metrics like an automobile dashboard. It has a few measurements that need to be monitored regularly and a few that can be observed with a lower frequency. Those metrics that are not vital to meeting the organization's strategic objectives can serve as "warning lights" for leaders.[15]

There is much interest in the topic of sustainability results, but there is less written on how the results must be linked to help the organization meet its strategic objectives: "Your measurements must focus on the past, present, and future, and be based on the needs of the customers, stakeholders and employees. Measuring everything is more damaging than measuring nothing—pinpointing the vital few key measures is the key to success."[16] An organization needs to focus on evaluating its current approach to measurement and then redesign the monitoring, measurement, analysis, and review process. Having bad data is worse than having no data at all.

All organizations collect some types of data. Trying to track everything for everyone is a big problem in the sustainability arena. It is important to locate and measure the right variables. But it is more important to monitor and measure the organization's processes correctly. All KPIs should be linked to the organization's success factors that will help meet the strategic objectives. These factors may be core values, technical competence of the workers, or marketing success. These links will help the leaders know how close the organization is to its strategic objectives.[17] Sustainability results must be linked to activities, processes, systems, products, and services associated with the operation of the organization.

An organization needs to have effective systems and processes for determining what data and information should be collected. There is also a need for efficient and effective processes to acquire, analyze, apply, and manage the information and knowledge. These measurements provide critical data about key processes, outputs, results, and other organizational outcomes. There is a need for information that is internal to the organization. The core practices include many of the following, whether formal or informal in nature[18]:

- Planning of data collection and linking it to the strategic planning
- Analyzing and interpreting data to learn and inform strategic planning
- Sharing the data among those who can use it to improve performance
- Ensuring data integrity (i.e., valid, relevant, timely, secure, and sound)

Some organizations use measures from the past. How did the costs today compare to the same location's costs a year ago? We do this when we look at our electricity bills. Measuring the current performance is critical to any organization, even though the moment it is measured, that result is in the past (i.e., lag measures). Past and present metrics are the easiest to come up with, because we typically have data on these types of measures and these results have actually happened. It is important to have lead measures in the measurement system (Chapter 16).

Selecting the right metrics or measures is actually much more than simply deciding what to measure. It is a key part of the organization's strategy for success. Selecting the wrong metrics will put a strain on the success of the organization. Many small businesses are using a generic "balanced scorecard" method so they can have fewer measures and still be able to drive success. Organizations are learning that success is about balance, not a dependence on any fixed set of measures.[19]

Like objectives, metrics need to be set at the highest levels of a hierarchical organization and followed down to all levels and functions. Metrics at one level should lead to metrics at the next level of the organization. Defining key performance measures in this manner ensures that there are no disconnects or inconsistencies in how the parent organization measures performance.

One way of reducing the number of measures to a reasonable number is to assign a weight to each individual measure in a family of metrics and develop an "index" that is an aggregate statistic. A number of national performance measurement programs enable organizations to score their results and then aggregate the scores to create an aggregate category (e.g., environmental impact). Combining multiple metrics into a single index is an excellent way of aggregating and simplifying reporting on performance.[20]

Strategic Measurement Model

While measurement may be easy, it is difficult to measure the right things and learn to ignore other interesting data that does not help the organization become more sustainable. There have been a number of different approaches on how to select and use metrics in an organization. Larger organizations and parent organizations use externally derived collections of sustainability metrics to guide the development of their strategy[21] (Figure 15.1).

The argument is made that sustainability reporting is a form of internal monitoring and management in order to have effective engagement with the stakeholders. This path has an organization focus on its impacts so that it can improve its operations in a manner that lessens those impacts. The organization's operations could be conducted with stewardship in mind, while creating the measurements for employee engagement. It is important not to see the sustainability report as being the sustainability program, but rather as a means of keeping track of meeting the strategic objectives over time.

It is possible to take a more internal view by looking at metrics from the internal perspective[22] (Figure 15.2). In this model, the organization determines what it stands for and its vision of the future. Obviously, its external impacts and stakeholder engagement will shape the vision. Next, the organization focuses on what it needs to do in order to maintain its competitive advantage, as well as its social license to operate. This would involve understanding its internal and external context. Usually, the organization will start by determining its lead indicators and determine how they can be measured

FIGURE 15.1
Sustainability reporting as a management process. (Adapted from GRI IFC, Getting more value out of sustainability reporting, 2010, https://www.globalreporting.org/resourcelibrary/Connecting-IFCs-Sustainability-Performance-Standards-GRI-Reporting-Framework.pdf.)

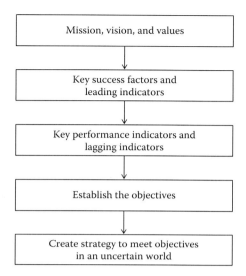

FIGURE 15.2
Strategic measurement model. (Adapted from Brown, M.G., *Keeping Score: Using the Right Metrics to Drive World-Class Performance*, Quality Resources, New York, 1996.)

as a means of assessing how effectively the lead indicators are working to influence the sustainability results (i.e., lag indicators). It is important to remember that results are merely the outcomes of performance. Results are not a direct measure of performance. It is important to create the results that are keeping score so that changes can be made in the support activity that are the focus of the lead indicators.

Objectives should be set only after there has been some experience with the determinations of the critical success factors. Often, there is a rush to set the strategic objectives. This leads to objectives that are surpassed with little effort or objectives that cannot be reasonably achieved. Neither of these extremes is helpful for driving an organization to meet its strategic objectives over the long term. Metrics are the tools used to monitor the progress to meeting responsibly established strategic objectives. This reminds us that the sustainability strategy should be prepared after the measurement system is already in place. One should never start with the strategy and derive measures that measure the progress on that strategy.

Roles of Monitoring and Measurement

When any kind of determination is needed, there is a need for monitoring and measurement. A relationship diagram highlights the respective roles

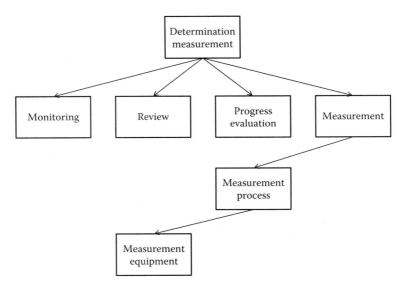

FIGURE 15.3
Measurement relationships diagram. (Adapted from ISO, Quality management systems—Fundamentals and vocabulary, ISO/DIS 9000, ISO, Geneva, 2014.)

of monitoring and measurement in the overall process (Figure 15.3). It is instructive to see how monitoring and measurement play different roles in the process of obtaining data for a determination[23]:

- *Characteristic*: Activity to find out one or more characteristics and their values
- *Review*: Determining the suitability, adequacy, and effectiveness of a product or service to achieve established objectives
- *Monitoring*: Determining the status of a system, process, or activity
- *Progress evaluation*: Assessment of progress made on achievement of the objectives
- *Measurement*: Process to determine a value
- *Measurement process*: Set of operations to determine the value of a quantity
- *Measuring equipment*: The measuring instrument, software, measurement standard, reference material, or auxiliary apparatus necessary to realize a measurable process

Note that monitoring can determine the status with observations or measurements. There are relationships between all these different practices as they work together to make important determinations in an organization.

> ### BOX 15.1   ESSENTIAL QUESTIONS ON SCOPING THE MONITORING AND MEASURING PROCESS
>
> Why does an organization monitor and measure to support its sustainability program?
>
> How does the sustainability reporting model differ from the strategic measurement model?
>
> What is the difference between monitoring and measurement? Why do you need both?

Monitoring and measurement are not conducted for the sole purpose of addressing the interests of stakeholders. The stakeholders are interested in an outcome and want evidence of what is actually happening in that regard.

In Chapter 16, there is more detail about what is involved in the processes of monitoring and measurement. We will also examine leading and lagging indicators that are associated with the monitoring and measurement processes.

Essential questions for scoping the monitoring and measuring process can be found in Box 15.1.

Endnotes

1. Hubbard, 2010.
2. Hubbard, 2010.
3. Thor, 1994.
4. Thor, 1994.
5. SAI Global, 2007.
6. SAI Global, 2007.
7. SAI Global, 2007.
8. SAI Global, 2007.
9. SAI Global, 2007.
10. Baldrige, 2013.
11. Hubbard, 2010.
12. Hubbard, 2010.
13. Hubbard, 2010.
14. Hubbard, 2010.
15. Brown, 1996.
16. Brown, 1996.
17. Brown, 1996.
18. SAI Global, 2007.

19. Brown, 1996.
20. Brown, 1996.
21. GRI IFC, 2010.
22. Brown, 1996.
23. ISO, 2014.

16

Monitoring and Measurement in Organizations

An organization needs effective processes and systems for determining what information and data should be collected to determine whether it is achieving its strategic objectives. *Monitoring* is used to determine the status of an activity, process, or system. *Measurement* is the process for determining a value or measure. The outputs from the monitoring are periodically reviewed to check the current situation, for changes in the work environment, conformance with organizational practices, and regulatory compliance. Monitoring, measurement, and review are activities that are used to determine the suitability and adequacy of processes and operations, and how the achievement of effectiveness and efficiency enable the organization to meet it objectives in an uncertain world.

Measurement produces results that are *lag indicators*. There are a number of widely used performance measurements that contain criteria that enable quantitative evaluation of organizational performance and that are applicable to the interests of stakeholders. These performance indicators are *leading indicators*. Their assessment criteria permit organizations to benchmark their processes, operations, products, and services with other organizations, as well as providing a useful "looking forward" sustainability perspective.

Monitoring

Monitoring involves the routine surveillance of actual organizational performance in order to create an accurate comparison with the expected or required performance. This activity involves continual checking or investigating, supervising, critically observing, and determining the process or system status to identify these changes, as well as changes in the context of the organization.[1] The main purpose of monitoring is to generate information needed to ensure that opportunities and threats are managed effectively. These effects of uncertainty and the operational controls and response may change over time. Those responsible for the embedded sustainability efforts need to be aware of the implications of any changes in the internal or external operating environment.

Monitoring requires a systematic approach that involves[2]

- Establishing a procedure for continual checking, supervising, critically observing, or otherwise determining the status of information or systems
- Developing a means of detecting change from what has been assumed or is expected
- Incorporation of the organization's performance indicators and the interests of the stakeholders
- Determining how resulting information can be captured, analyzed, reported, considered, and acted upon
- Providing the necessary resources and expertise for these activities
- Allocating responsibilities for various risk management and uncertainty monitoring activities and incorporating those responsibilities in the member's or employee's performance review criteria

Monitoring is an aspect of effective *governance* to make sure that uncertainty is managed effectively. The selection of things to be monitored should be focused on the significant opportunities and threats.

Next, it is important to determine *what* needs to be monitored. Here are some typical monitoring examples:

- Key characteristics of operations that have significant impacts on the environment, society, or the local economy and meeting the organization's three responsibilities
- Processes in place that ensure compliance with legal and other compliance obligations
- Key characteristics associated with the interests of stakeholders
- Significant organizational opportunities and threats
- Operational controls associated with the system of management
- Value chain controls associated with suppliers, customers, and stakeholders
- Interactions found in the value chain model
- Determinants of progress toward the organization's strategic objectives

The organization needs to determine a method that will be used for monitoring that is consistent with the measurement method selected. It is also important to ensure that any monitoring device (e.g., sensor) is calibrated and maintained as required by the manufacturer's specifications.

Documented information is needed to review the results of the monitoring activity. Included in these records are the following:[3]

- Information on how monitoring decisions were made
- Information regarding the interests of stakeholders
- Controlled records for any reporting that is required
- Assistance to the organization in the review of its activities, processes, and systems
- A demonstration of whether the process has been conducted in a planned and systematic manner
- Enabling of information about the process to be shared through engagement with internal and external stakeholders
- Provision of objective evidence for internal and external process audits

Leaders need to establish and maintain processes for monitoring and collecting and managing the information obtained from the monitoring activities.

Measurement

All organizations collect different types of data. Trying to track too much information is a big problem in the sustainability arena. It is important that the organization measure the activity associated with the significant opportunities and threats, the effectiveness of the processes, and the efficiency of the activities, processes, and systems. These measurements provide critical data about key processes, outputs, results, and other outcomes. Measurement focuses on the past, present, and *future* (i.e., looking forward results) and may be based on the needs of employees, management, customers, and stakeholders. The key to a successful measurement program is pinpointing the vital few key measures.[4]

Measurement is a process for determining a value.[5] Measurements should be either quantitative or semiquantitative. It is important that measurements be conducted under controlled conditions with appropriate processes and procedures for ensuring the *validity* and *traceability* of the results. These processes include adequate calibration and verification of monitoring and measurement equipment, use of qualified people, and use of suitable methods and quality control procedures. Measuring equipment should be calibrated or verified at specified intervals or, prior to use, against measurement standards traceable to international or national reference standards. If the standards do not exist for a particular measurement, the basis used for the calibration and use of the equipment must be recorded. Written procedures for conducting measurement and monitoring are necessary to provide consistency in measurements and enhance the reliability of the data that is produced.

Leaders need to assess the progress of the organization in achieving planned results against its strategic objectives, at all levels and in all relevant processes and functions. A measurement and analysis process is used to gather and provide the information necessary for performance evaluations and effective decision making. The selection of appropriate key performance indicators (KPIs) and measurement methodology is critical to the success of the measurement and analysis process.[6]

Lag Indicators

Lag indicators are measures that focus on the outcomes or results of planned initiatives at the end of a time period. Results or KPIs largely indicate historical performance. There are large numbers of targeted sustainability indicators and KPIs available in the literature. However, the challenge in using them includes the fact that these results do not reflect current activities. They lack predictive power. There is a caution in the investment world that states, "Past performance offers no indication of future results." Past performance is useful in valuing the sustainability program only as far as it is indicative of what is to come in the future since sustainability is always defined as happening over the long term. Many sustainability practitioners regard past performance as being able to at least provide an idea of the direction in which the organization is headed. However, a performance trend is not considered to be a leading indicator.

Although sustainability practitioners and stakeholders tend to focus on results (lag indicators), these lagging indicators may not provide enough information to guide actions and ensure success.[7] There are some good reasons why lag indicators may not always be sufficient for measuring sustainability:

- Lag indicators often provide information too late to enable sufficient response.
- Outcomes are the result of many factors. Lag indicators may tell you how well you are doing, but may not give information as to why something is happening and where to focus any corrective actions that may be necessary to improve performance.
- When the outcome rates are low, there may not be sufficient information in the measures to provide adequate feedback for effective management of the process.
- When the outcomes are so threatening that you cannot wait for it to happen before the process is upset.
- Lag indicators may fail to reveal latent threats that have a significant potential to result in disaster.

Lag measures tell you if you are about your success in terms of achieving the strategic objective as a "work in progress" measurement. It is difficult to do anything about a lag measure once it is reported because lag measures have already happened when you measure them. You are always referring to the past when talking about a result (lag measure). They are like a photograph—activity frozen in a past point in time. Yet despite the problem of not possessing a looking forward capability, lag measures are believed to be a sign of succeeding with a strategic objective by getting the results to demonstrate this achievement.

Lead Indicators

A lead performance indicator is something that provides information that helps the user respond to changing circumstances and take actions to achieve desired outcomes or avoid unwanted outcomes. The role of a lead indicator is to help organizations improve future performance by promoting action to correct or avoid potential weaknesses without waiting for something to go wrong. The ability to guide actions to influence future performance is an important characteristic of lead indicators.

There are two forms of lead indicators. One form of a lead indicator involves providing information about the current situation that will affect future outcomes. A second form of lead indicator involves the measurement of processes involving people that can make a process more effective and connected to meeting the organization's strategic objectives.

A good example of the former lead indicator is when an economist uses certain lag indicators, such as "new housing starts," that foretell of the demand for goods necessary to build and furnish those homes. This reinforces the tendency of finding patterns with results (lag indicators).

The latter form of leading indicator involves the measures of activities connected to meeting the organization's strategic objectives. These measures are within the control of the organization. This lead measure (e.g., leadership) is different since it can foretell the result. You can recognize lead measures because of the following characteristics[8]:

- A lead measure is *predictive*—if the lead measure changes, you can predict that the lag measures associated with the lead measure will also change.
- A lead measure is *influenceable*—it can be influenced by the organization that seeks to manage the activities connected to meeting the organization's objectives.

For many people, lead measures seem to be counterintuitive. For example, we know that leaders and sustainability practitioners always focus on the lag indicators, even though it is not possible to act on a lag measure since it is always in the past. There is a feeling that lead measures are hard to keep track of since they measure behaviors and behavior change. Finally, too many people, lead measures often look too simple. Since lead indicators focus on behaviors such as leadership, employee, and stakeholder engagement, those that are not involved with these activities directly think that they are not as important as a measure for sustainability.[9]

In the fields of organizational development and risk management, we are focused on behaviors (i.e., positive effects) that help the organization achieve its strategic objectives. The information on lead indicators makes a difference and enables the organization to close the gap between what the organization should do and what it is actually doing. Risk managers often describe lag indicators as driving an automobile with the front windshield painted black, moving forward by navigating with the three rearview mirrors. So how does an organization prepare for creating lead indicators to drive the lag indicators over the long term?

Lead Indicators and the Process

Processes always present some challenges since people are focused on results[10]:

- Is the process producing good results?
- Are people carefully following the process?
- Is the process effective with efficient operations?

The process focus element of performance frameworks looks for leverage points in the process. These are those critical steps in the process where performance can falter. Lead indicators are linked to these points to improve the results using leadership, strategic planning, employee engagement, stakeholder engagement, information and knowledge management, and process focus.

The trick is to use the leverage of the people support processes in the Porter value chain model to trigger the improvement of the results (lag measures) over time. The approach, deployment, assessment, and refinement (ADAR) approach is used to accomplish this. Once you have a goal to meet an objective, there is a need to look to the performance area that would be responsible for providing the leverage factor, as shown in Figure 16.1. The process focus category in the performance framework can be used to guide

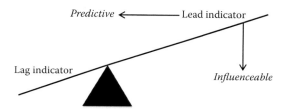

FIGURE 16.1
Acting on lead indicators. (Adapted from McChesney, C., et al., *The 4 Disciplines of Execution*, Simon & Schuster, London, 2012.)

the development of the approach. This is followed by a plan to deploy the process focus approach to guide the improvement of the lead measure. When this combination works well, the lead measure will be predictive and influence the lag measure. The lag measures that are attached to each lead measure should vary in the same manner.

Performance Frameworks

Performance frameworks have been used in approximately 70 countries. These frameworks provide the foundation for what are referred to as national quality award programs. Performance excellence programs were introduced in the late 1980s to spur the competitive advantage of businesses in those countries. In the United States, the Baldrige Performance Excellence Program has been in continuous use since 1987. This program has developed a set of criteria for performance excellence to be used by participating organizations, which include large and small businesses, government, nonprofits, schools, and healthcare facilities. There is a scoring method that enables the assessors to score the applications and grant the award to those who perform best.

Many organizations have learned to use the performance frameworks as a means to improve their competitiveness without applying for the award. Because the frameworks are updated on a regular basis to reflect changes in the best practice for each of the categories, they provide a useful means for looking forward and driving performance.[11] This, in turn, helps improve the results. The developers of these frameworks noted that *results are merely the outcome of performance* and do not measure performance directly. What is different with the performance frameworks is their focus on driving performance in each of the following areas:

- Leadership
- Strategic planning
- People and employee engagement

- External stakeholder engagement (including customers)
- Information and knowledge management
- Process focus

Each of these areas can be scored and act as a lead indicator. Each lead indicator is associated with results (lag indicators) to see if the lead indicators are influencing and predicting the lag indicators. By measuring the inputs to a process, lead indicators can complement the use of results (lag indicators) and compensate for some of their shortcomings. Lead indicators of this nature can also be used to monitor the effectiveness of process controls giving advance warning of any developing weaknesses before problems occur.

Using a Performance Framework

A performance framework contains information on what is recognized as a best practice in each of the categories.[12] To keep current, the framework must be continually modified to reflect current best practice. Any organization can use these descriptions to benchmark against and make corresponding improvements in its processes and operations.

Three documents are prepared by the organization for each of the performance categories selected to help guide the efforts to attain the level of performance that is sought. These documents include an *approach*, the means of *deploying* that approach, and the manner in which the organization will *assess* and *refine* the approach and the deployment of the course of the implementation period—usually one year.[13]

A description of how to prepare each of these documents is provided below. This is followed by a description on how to score them so they can act as a semiquantitative measurement. These measurements will become the lead indicators for the sustainability program.

Approach

To create semiquantitative lead indicators that drive performance for each of the sustainability indicators, an organization needs to create its strategy. How will it address the various human support processes listed above to improve its current operating state? The approach of the ADAR cycle is basically the planning stage of the sustainability process. An organization must establish the strategic and operational objectives and the processes necessary to deliver sustainability results or KPIs.

The approach identifies the organization's intent to become sustainable and to measure its progress. In other words, the approach is the way by which something is made to happen.[14] It consists of processes and actions that take place within a framework of principles and policies.

In order to create the approach for each category, it is important to answer the following questions[15]:

- What is the organization seeking to achieve for each category? What is its intent?
- What strategic and operational objectives and worker goals and action plans have been established?
- What strategies, governance, and processes have been developed to achieve the intent? How were these items chosen by the leadership of the organization?
- What indicators or KPIs have been designed to track the progress of the organization's performance?
- How does the approach provided align with the strategic objectives established by the organization?

The approach should not only specify *what* an organization is planning to accomplish, but also include the reason *why* it is planning to do this.[16] In order for the approach to be sound, the organization should consider many of the following[17]:

- Interests of the stakeholders as determined in the engagement process
- Alignment with the organization's strategic objectives and strategy
- Appropriateness for the organization's internal and external context
- Links to other approaches, if appropriate
- Sustainability principles
- Whether it enables the organization's approach to continual improvement, innovation, and learning
- Consideration for the scoring matrix used to create the semiquantitative score

By addressing the information provided above, the approach will ensure that an organization's sustainability strategy correlates with a planned and strategic cycle of improvement. The stronger the adherence to these items, the higher the score that will be given to the approach. When items are not addressed affirmatively, the score will be lower. A weak approach typically addresses what was to be accomplished, not how or why. A well-defined approach that is supported with some benchmarking information tends to score higher using the scoring matrix.

Deployment

The deployment of the approach identifies how an organization plans to implement those processes and activities to create this lead indicator component. Successful deployment involves embedding the approach into the activity and full acceptance of the process steps and activities found in the approach. The deployment goal is to achieve the intent that was stated in the approach.

To successfully implement the approach, the organization should consider preparing a draft action plan.[18] Essentially, the action plan heading should describe the area being addressed (i.e., leadership, strategic planning employee engagement, and so forth). It should also include the purpose of the action or the intent, and state the expected benefits. Next, each of the tasks should be outlined assigning them to individuals, providing a timeline, and highlighting the anticipated performance improvements required for completing the item. The tasks will be the key process steps and procedures outlined in the approach. Usually, the first task of the action plan should describe what is being measured during the deployment and the last step should recap the lessons learned by the employees and leaders as a result of the deployment of the approach.

To assess the success of the deployment (i.e., implementation plan), organizations should be able to describe the following:

- How have the strategies, processes, and activities been executed with the employees? What is the effectiveness of their implementation throughout the organization? To what extent have they been accepted and embedded as part of normal operations?
- Has the approach been implemented across all relevant organizational areas to its full potential and capability? Do the planned implementation activities support operations and sustainability improvement?
- Has the deployment of the approach been executed in a timely fashion and structured in a way that enables it to adapt to changes in the internal and external operating environment?
- Is the deployment of the approach achieving the planned benefits?
- Is the deployment of the approach understood and accepted by the internal and external stakeholders?

A written action plan will facilitate the evaluation of the deployment and enable the organization to address each of the questions that are posed during the scoring of each approach and deployment.

Assessment and Refinement

Assessment and refinement provides a process of improvement that enables the approach and deployment statements to be reviewed and modified to achieve the best possible results. Through monitoring and analysis of these results, the level of effectiveness of the sustainability program can be measured. An efficient and highly rated organization will conduct learning activities in the organization in order to prioritize, plan, and promote creativity, innovation, and improvement. This leads to three categories of an assessment and refinement: measurement, learning and creativity, and innovation and improvement.

Measurement

The approach and deployment must be regularly monitored and measured to ensure efficiency and the implementation of processes. Monitoring and measurement enable the organization to indicate that the results indeed stem from the deployed approach. The organization should be able to answer the following questions to assist in the assessment and refinement:

- What processes are in place to review the appropriateness and effectiveness of the approach and deployment?
- How do the results of the approach and deployment compare with benchmarks or previous levels of performance?
- How do you communicate, interpret, and use these results?

Learning and Creativity

In a learning stage, an organization identifies the best internal and external practices so that it will be able to recognize improvement opportunities in its sustainability program (Chapter 3). Through the use of creative measures, the organization can then refine the approach and deployment by encouraging continual improvement and innovation. Answers to the following questions are addressed by learning and creativity:

- What have you learned?
- How have you captured this learning?
- How can you use the learning to improve the approach and deployment?

Improvement

Finally, the organization should use the results from the measurement, learning, and creativity to plan and implement the identified improvements (Chapter 19). These refinements and innovations should be shared across all relevant approaches for consistency and continual improvement. At this point in the process, the organization can state how it has used the learning to improve the approach and deployment.

Scoring the ADAR Categories

Performance frameworks score the ADAR categories using a special scoring matrix (Table 16.1). The approach is analyzed to determine its soundness in providing a clear rationale, defined processes, and a focus on stakeholder interests. It should have strong ties to the sustainability and the organization's efficacious strategy. The deployment is analyzed to determine whether it is systematic and effectively implemented. It is important to ask whether it

TABLE 16.1

Performance Scoring Matrix

Score	Approach	Deployment	Assessment and Refinement
0	No evidence that the approach has been considered and there is a reactive attitude to performance.	Anecdotal information on how the approach has been deployed. No evidence that a deployment action plan has been used.	Anecdotal information on how the approach and deployment have been addressed. No results shown and no improvement activities in place.
1–2			
3–4			
5	Approach has a rationale and is proactive with defined processes. May not address strategy and recognition of stakeholder interests. No improvement from assessment and refinement.	Approach is applied to many areas and activities. Approach is becoming part of operations and planning with formal action plans and accountability. Evidence that results are caused by approach in some areas.	Positive trends in many areas. Results are comparable with benchmarked findings in many areas. Evidence that results are caused by the approach and deployment is subject to ad hoc review. Evidence that some improvements are implemented.
6–7			
8–9			
10	Approach is accepted as best practice and is benchmarked by other organizations seeking to improve. All items in the guidance are evident.	Approach is deployed to all areas and activities. Approach is totally integrated into normal operations and planning.	Results are clearly caused by the approach in all areas. There is a proactive system for regular review and improvement of the approach and deployment.

Source: Adapted from EFQM, EFQM excellence model, EFQM, Brussels, 2010.

was conducted on time and in a structured way. The scoring of the approach and refinement is threefold, beginning with the measuring of the effectiveness and efficiency of the approach and deployment, and then learning and creativity, and improvement and innovation.

When scoring each ADAR category, the assessor should start by reading the information in the middle row (i.e., 5 points). If the evaluation is deemed to be better than that description, the assessor looks at the low and high side of the next higher row (i.e., 6–7 points). On the other hand, if the evaluation is deemed to be less positive than the middle row, the assessor looks to the wording in the next lower row (i.e., 3–4 points). The movement continues until the assessor is comfortable with the evaluation in the scoring matrix as a reasonable explanation of what was observed. The assessor scores each of the ADAR categories in order, from right to left in the scoring matrix. Typically, the deployment lags the approach and the score is typically at least 1 point lower. The assessment and refinement lags the approach and the deployment, so the score is typically 1 point lower than the deployment score. It takes a little practice to use the scoring matrix.

Remember that the leadership is using the scoring sheet when preparing the approach, deployment, assessment, and refinement documents. Some organizations start with a modest approach in the range of 5 points. If successfully put into use, the approach will generally score between 4 and 6 points. The scores get higher when the organization has more experience with this process. As mentioned, there are typically up to six processes that are implemented as leading indicators, as found in the top section of the Porter value chain diagram. No organization is equally proficient in the ADAR of the six lead indicators. You may see some approaches in the 8–9 scoring range and some in the 4–5 scoring range. As the best practices improve, it becomes

BOX 16.1   ESSENTIAL QUESTIONS ON MONITORING AND MEASUREMENT

How do monitoring and measurement work together to help an organization create and maintain effective processes and efficient operations?

How do you define the manner in which lagging and leading indicators work together within the monitoring and measurement system of an organization?

How can you best describe each of the people activities at the top of the value chain model for a particular organization in advance of scoring each activity using the ADAR method?

How do the ADAR scoring categories enable an organization to quantify the effectiveness of the categories at the top of the value chain model?

more difficult to get the same score as the year before with the same amount of effort. On the contrary, the organization must improve its ADAR methods to maintain its score as a result of the improvement in best practice.

Essential questions for monitoring and monitoring can be found in Box 16.1.

In Chapter 17, the use of the processes for measuring lead and lag indicators is covered in more detail.

Endnotes

1. ISO, 2013.
2. AS/NZS, 2013.
3. ISO, 2009.
4. Brown, 1996.
5. ISO, 2008.
6. ISO, 2009b.
7. Step Change in Safety, n.d.
8. McChesney et al., 2012.
9. McChesney et al., 2012.
10. McChesney et al., 2012.
11. Pojasek, 2007.
12. Pojasek and Hollist, 2011.
13. EFQM, 2010.
14. EFQM, 2010.
15. SAI Global, 2007.
16. EFQM, 2010.
17. EFQM, 2003.
18. Pojasek, 1997.

17

Transparency, Accountability, and Sustainability Performance

At the organization level, information and stories are transmitted to the parent company for inclusion in a sustainability report. When operating separately, the monitoring and measurement activities that take place at the organizational level let the leader know whether there are effective processes and efficient operations. Some of this data and information is shared within the stakeholder engagement program or with customers that request information.

No matter where the sustainability program is located, attention must be paid to the concepts of transparency and accountability. In any organization, *transparency* provides a sense of openness about decisions and activities that affect the environment, the community, and the economy, as well as a willingness to communicate this information in a clear, accurate, timely, honest, and complete manner.[1] *Accountability* is all about being responsible for decisions and activities to the organization's governance, legal authorities, financial authorities, and more broadly, the stakeholders.[2] The *performance* of the organization is measured relative to the achievement of its strategic, operational, and tactical objectives. Organizations seek levels of performance that deliver ever-improving value to all stakeholders, thereby contributing to organizational sustainability.[3] The quest is always for effective processes, efficient operations, and efficacious strategy. Having examined the monitoring and measurement tasks, we will now explore some of the outcomes of these efforts, and perhaps discover even more sustainability measures or disclosures that need to be made to different stakeholders.

Transparency

In the practice of sustainability, an organization is expected to be transparent in its decisions and activities that have impacts on the environment, society, or the economy. As such, a local organization is expected to disclose in a clear, accurate, and complete manner, and to a reasonable and sufficient degree, the policies, decisions, and activities for which it is responsible. This disclosure should include information on the known and likely impacts that the organization may have on the local environment, the people in the community, and its shared values.

Until the organization has secured its "social license to operate," stakeholders may expect the organization's information to be readily available, directly accessible, and understandable to those who have been or may be affected in significant ways. Information should be timely and factual and presented in a clear and objective manner. This should enable the outside stakeholders and the organization to engage in discussions about the impact that the organization's decisions and activities have on their respective interests.[4] Under the best of conditions, much of the disclosure takes place within an established stakeholder engagement process.

The principle of transparency does not require that proprietary information be included in the disclosures. Furthermore, transparency does not involve providing information that is privileged or would breach legal, commercial, security, or personal privacy obligations.[5] Often, there is a contentious time in the process of stakeholder engagement when the organization has not built a sufficient level of trust in the view of the stakeholders. The information is used to determine the organization's performance until trust is established.

Many organizations find themselves in a situation where the more information that they provide to the stakeholders, the more information that the stakeholders may demand to have provided. This can be remedied by using a risk assessment method to examine the failure modes (i.e., the impacts) and find a way to address the root cause of the situation referred to as the "impact" so that the stakeholder engagement process can focus more on whether the organization is meeting its strategic objectives and whether these objectives will have positive effects for the stakeholders as well. Failure mode effects and analysis has been adapted to manage all three responsibilities of sustainability.[6]

There are a number of other things an organization should consider when seeking to become more transparent[7]:

- Explaining the purpose, nature, and location of its activities at the community level
- Identifying people with a controlling interest that are involved in the organization
- Manner in which decisions are made, implemented, and reviewed, including the definition of the roles, responsibilities, accountabilities, and authorities across the different functions in the organization
- Standards and criteria against which the organization evaluates its own performance relating to sustainability
- Performance on relevant and significant interests of the stakeholders in the area of sustainability
- Sources, amounts, and application of its funds

- Known and likely impacts of its decisions and activities on its stake-holders, society, the economy, and the environment
- Stakeholders and the criteria and procedures used to identify, select, and engage them

The use of such an extensive list would be difficult for most organizations at the community level. For large hierarchical organizations, people would have to be hired just for the purpose of fulfilling the demand for transparency in a sustainability program.

Accountability

In the practice of sustainability, an organization is expected to be accountable for its impacts on the environment, society, and the economy. This suggests that an organization should accept appropriate scrutiny and a duty to respond to this scrutiny.[8] Sometimes this concept moves away from engagement with stakeholders and assumes that there will be watchdog organizations that are looking to see what organizations are creating impacts. Would not this definition be more consistent with the transparency definition if the accountability were to engage stakeholders directly and discuss the information that was provided in a transparent manner?

It is well accepted that accountability involves an obligation of the organization to be answerable to legal authorities with regard to laws and regulations. However, the concept of accountability is expanded by sustainability to include a similar obligation for an organization's overall impact of decisions and activities on the environment, society, and the economy to those affected by its decisions and activities, as well as to society in general.[9] Accountability also includes accepting responsibility where wrongdoing has occurred, taking the appropriate measures to remedy the wrongdoing, and taking action to prevent it from happening again.[10]

Many believe that organizations should base their behavior on standards, guidelines, or rules of conduct that are in accordance with accepted principles of doing the right thing or practicing good conduct in the context of specific situations, even when these situations are challenging.[11] Finally, organizations are expected to respect, consider, and respond to the interests of their stakeholders.[12] The process of engagement helps establish an understanding where both parties would be expected to do the same. This is one of the outcomes when the organization is granted its social license to operate. It is easy to see that we have a long way to go in how these very important principles are used to get parties to engage with each other just as the word is meant to convey.

Organizational Basis for Disclosures

An organization seeks to meet its strategic objectives in an uncertain world. Let us use a concept that is often used for businesses to look at how organizations seek to disclose information on what they are doing to achieve these objectives. There is a continuum in all organizations that moves from the mission, to strategic objectives, to strategy, to execution and the determination of the organization's performance. Each organization should pay attention to the "voice of the customer and stakeholder" in each step along the way. The outcomes of the well-executed continuum include engaged stakeholders (internal and external to the organization) and effective processes and efficient operations. Within a sustainability program, the internal and external performance is usually disclosed both within the organization and external to the organization. If there is an emphasis on engagement of stakeholders throughout the continuum, the disclosure of performance is generally not found to be contentious. It is rather a statement on how well the organization and stakeholders are engaged.

The economic disclosures come from the realization of the first principle of sustainability management:

> Our organization exists to create and protect value for our members, employees, customers, and stakeholders.

Sustainability contributes to the demonstrable achievement of objectives and improvement of organizational performance.

The society perspective represents the successful realization of stakeholder engagement—to both the internal organization and the external stakeholders and society in general. There are different levels of engagement that the organization must navigate in this perspective[13]:

- Identify stakeholders during the scanning of the internal and external operating environments
- Begin communicating with the stakeholders
- Seek engagement with the stakeholders
- Grow relationships with stakeholders

Every organization has a broad range of stakeholders.

The organization's operations can be divided into a number of processes that are important in the environmental perspective[14]:

- Develop and sustain the organization's niche in important value chains
- Produce products and deliver services in an efficient manner

- Distribute and deliver products and services to stakeholders (e.g., customers)
- Manage the organization's risks associated with these processes

The process and activity of the organization is focused on the effective and efficient conservation of resources used in services and products that can pose threats to the environment if the organization does not manage its operations, customers and stakeholders, and three responsibilities, and take advantage of innovation associated with its performance improvement.

The learning and growth perspective is clearly driven by the sustainability management principles (Chapter 2) as they are applied to the sustainability program. Many of the value chain illustrations benefit from the judicious application of these sustainability principles.

Context and Performance Measurement

When an organization has a sustainability program with an *inward-focused perspective*, it is dealing with stewardship of the three responsibilities of sustainability, value chain management, quality management, availability of people that can accept the organization's culture, and a system of management. There is a focus on efficiency, growth, financial structure, and other operational attributes. Much of the monitoring and measurement is focused on meeting its strategic objectives over the long term and successful engagement with the stakeholders. Sustainability can be embedded in the way the organization operates every day.

When an organization has a sustainability program with an *outward-focused perspective*, it begins to demonstrate its awareness on how environmental, social, and economic factors affect the operation of the organization. There is a focus on meeting objectives by managing the positive and negative effects (i.e., opportunities and threats) of uncertainty. It will begin to quantify the threats and opportunities related to environmental, social, and economic factors. The organization will start to quantify the threats and opportunities related to its three sustainability responsibilities. This begins to demonstrate through stakeholder engagement that sustainability is part of the internal strategy and will lead to performance that recognizes both the internal and external context.

Trust is what sustainability disclosure is all about. Transparency and accountability help the organization establish credibility. They demonstrate the effectiveness of stakeholder engagement both to the internal organization and with external stakeholders. The organization begins to see transparency as allowing the will of the stakeholders to play its part in monitoring the trustworthiness (i.e., validity) of the sustainability disclosures. Organizations with an internal-only focus are much less likely to share information from the management of the internal context. The major disconnect between the three-responsibility performance and internally focused performance is the latter's focus on impacts and regulations, which are seen as a cost center instead of a value center. Many of the sustainability measures currently in use are not designed to optimize three-responsibility performance or to understand the long-term impacts of organizational decisions. Most of the metrics currently in use focus on one responsibility and are internally focused.

Risk and Performance Measurement

The corporate sustainability practitioners use a sustainability management method referred to as "materiality." This is a financial term that is being applied to all the organization's nonfinancial metrics. The materiality matrices in use today have the external context (interest to the stakeholders) on the y-axis and the internal context (interest to the organization and its operations) on the x-axis.

In the materiality determination, the y-axis on the grid has one of the following labels:

- Importance to external stakeholders
- Community, neighborhood, or society concern

These concerns constitute the external context of the organization.

The x-axis or the materiality plot consists of one of the following different labels:

- Likelihood of an impact to an organization
- Influence on the organization's objectives
- Current or potential impact
- Significance to the organization

These concerns constitute the internal context of the operation.

In order to engage with stakeholders and improve the operation, it seems to make more sense to discuss the uncertainty (both opportunities and threats) of the organization and how uncertainty is being addressed through the uncertainty assessment and response efforts. The metrics would include the lead indicators to show how management is dealing with the effects of uncertainty through the efforts of the people support activities found in the top half of the value chain model. The lag indicators or the results of those efforts are reflected in the suppliers, inputs, process, outputs, and customers (SIPOC) diagram at the bottom half of the value chain model. Just like the monitoring and measurement done with the process, it is best to select a few very important areas to focus on. By using the uncertainty assessment, an organization can talk about the approach to identify the effects of uncertainty and can then show how it plans to manage the significant opportunities and threats, along with the risk of achieving its strategic objective.

This is a more effective approach to sharing of sustainability information within the engagement of stakeholders process as long as the organization examines all its significant opportunities and threats, rather than selecting from an established list of sustainability results (i.e., lag indicators) and putting them on a materiality plot. The organization's reporting should involve presenting a clear picture of how dealing with the effects of uncertainty can help improve its ability to meet its strategic objectives. By focusing on the strategic objectives, it is possible to manage the expectations of the organization and drive the optimization of the results that are needed to demonstrate progress. Managing the results will involve balancing the response to the internal and external contexts by dealing with the significant opportunities and threats.

Measuring the Lag Indicators

All organizations analyze and evaluate appropriate data and information from their monitoring and measurement activities. These results (lag indicators) are used to evaluate many of the following[15]:

- Conformity of an organization's products and services
- Degree of customer satisfaction
- Performance and effectiveness of the system of management
- Degree to which the strategy is efficacious
- Effectiveness of actions taken to address the effects of uncertainty
- Performance of the value chain suppliers
- Need for improvement, innovation, and learning

Organizations may receive requests for specific results (i.e., lag indicators) from parent organizations as part of the sustainability reporting effort or from suppliers, banks, and other organizations that are dealt with on a regular basis. If the requested results are not currently collected as part of the monitoring and measurement effort, it is useful to engage the parties in a discussion of the process used to gather these results. The organization should seek to

- Measure aspects of the value chain process that create value
- Measure the activity outcomes that set the organization apart (i.e., competitive advantage)
- Measure what other similar organizations are measuring since this ensures that focused metrics are harmonized

However, the measures need to be mindful of stating the creation of value. The methods used by the organization to monitor and measure, analyze, and evaluate these results should ensure that[16]

- The timing of monitoring and measurement is coordinated with the need for analysis and evaluation of the results
- The results of monitoring and measurement are reliable, reproducible, and traceable
- The analysis and evaluation are reliable and reproducible, enabling the organization to report trends

Each of the lag indicators can be judged for *significance* on the basis of the following[17]:

- Is the result important or significant to the organization?
- Is the result associated with an action plan already in use?
- Is the result regularly tracked with the organization's leaders following its trends?
- Does the organization actively benchmark this result with others?

Positive answers to these questions show that this process of tracking results is improved by the attention given to the performance management process.

By engaging external stakeholders in this process, it should be possible to provide them with results that are more useful while not creating an extra effort within the organization. The obvious exception to this is the information that is needed to measure the regulatory compliance, along with other similar contractual obligations.

What about the accountability associated with the engagement with other external stakeholders associated with a parent organization? There may be requests for results that are not routinely collected. It is important that these requests be discussed within the engagement process since an investment will be necessary to collect that information. The external stakeholder needs to understand the full mechanism of monitoring and measurement, while the organization needs to understand whether an interaction with the external stakeholders requires such measurement.

Within the organization, goals and action plans are set in the lowest levels of the operations. It is important to have activities where the people involved only have one or two goals. The success of the action plan is based on the ability of the people that work with a process to meet these goals. Remember that the lag indicators are the tracing measurements used to determine whether the goals are being met. It is important to keep score. The strategic objectives are achieved through successfully meeting the operational objectives. As presented in Chapter 7, this process of requesting results is cascaded all the way to the bottom of the organization. Results must be kept at each level in the operations and then in the organization as a whole. Since the connection between levels is created at the initiation of the scoring dashboard, it is important that each level only views what is important to them. The leaders of the organization will manage the progress for achieving the strategic objectives and will communicate that information to the entire organization and external stakeholders—all at the appropriate time.

The results of the sustainability program can also be scored using the four attributes of lag indicators mentioned earlier. It is possible to score the 15 or 20 key performance indicators (KPIs) or other results and then aggregate those scores to a single lag indicator score. More information will be provided after looking at the lead indicators.

Designating the Lead Indicators

Based on the available performance frameworks, it is possible to create a number of leading indicators representing the support that people provide in the top half of the value chain model. Here is an example of these lead indicators:

- The leader embraces accountability for the organization's sustainable success.
- The organization's strategic plan is instrumental in mainstreaming sustainability.
- Engaged employees embrace sustainability as part of their daily work.
- Engaged stakeholders provide value through sense making and process improvement.
- The organization participates in partnerships when collaboration can further improve sustainability.
- A sustainable organization is created through effective processes and efficient operations.

Every organization creates a description of how these lead indicators are realized. Examples of these descriptions[18] can be found in the next sections.

Leadership

Excellent leaders ensure the organization's mission, vision, principles, values, and ethics reflect a sustainability culture that they role model and reinforce with the internal stakeholders to enable them to contribute to the organization's strategic objectives.

Leaders should define, monitor, review, and drive the improvement of the organization's system of management and performance to ensure that it addresses current and future environmental, social, and economic.

Leaders should also engage directly with external stakeholders in support society in their community by participating in capacity-building activities. In particular, they foster equal opportunity, the environmental quality, education, and health, and encourage well-being among community stakeholders by minimizing the adverse impacts of their products, services, systems, and processes.

Leaders ensure that the organization is flexible and manages change effectively, taking into account the organization's sustainability commitments and its legal, ethical, environmental, social, and economic responsibilities.

Strategic Planning Process

The organizational governance embeds sustainability into its policies, strategy, and day-to-day activities based on understanding the interests of the stakeholders in their external operating environment.

The organization's strategy is based on addressing the challenges faced by its internal performance and capabilities as benchmarked against the performance of competitors and "best-in-class" organizations for their environmental, social, and environmental impacts.

The organization's strategy and supporting policies are developed, reviewed, and updated in a manner consistent with its mission and vision and the organization's strategy for sustainability.

The organization's strategy and supporting policies are reviewed within the stakeholder engagement process and communicated with stakeholders in the community to raise awareness about sustainability. The strategy and supporting policies are communicated, implemented, and monitored by the organizational governance.

Employee Engagement Process

The organization manages, develops, and releases the full potential of its people at individual, team, and organizational levels, including involving and empowering them in discussions on sustainability and related activities and planning. This helps secure support of the organizational strategy and supporting policies.

People in the organization are encouraged and supported in developing their knowledge and capabilities.

People are aligned, involved, and empowered to meet the strategic objectives and the sustainability policy as part of what they do every day.

People must communicate effectively throughout the organization.

People are rewarded, recognized, and cared for within the organization.

External Stakeholder Engagement Process

The organization will identify its stakeholders in its process of scanning its external operating environment and seek to engage with them to help manage uncertainty in line with the organization's sustainability program.

The organization will seek to obtain and maintain the social license to operate that is granted by the external stakeholders in the local community by sharing interests and building relationships.

The organization will design and tailor processes for building and managing customer and stakeholder relationships to suit markets with the aim of acquiring new customers, retaining existing customers, and enveloping new market opportunities in line with the organization's sustainability program.

Finally, the organization will measure customer satisfaction and loyalty; compare the results with those of the competitors and use the information to improve internal processes, products, and services; and deliver increasing value for customers, markets, and other external stakeholders.

Partnerships and Resources Process

Partners and suppliers are managed for sustainable benefit by building relationships based on mutual trust, respect, and openness.

Finances are managed to secure sustainable success by developing financial strategies to support the organization's sustainability programs strategy.

Buildings, assets, infrastructure, materials, and natural resources are managed in a sustainable way.

Technology is managed by involving internal stakeholders and other stakeholders in the development and deployment of new technologies to maximize the benefits.

Information and knowledge are managed to support effective decision making and sense making, and to build the organization's capability to continually improve within its sustainability program.

Using Effective Processes and Efficient Operations

Processes are designed and managed to optimize stakeholder value, clearly linking the outcome measures to the organization's strategic objectives.

Products and services are developed to create optimum value for the customer while taking into account any impact the products and services may have on environmental, social, and economic responsibilities in the sustainability program.

Products and services are produced, delivered, and managed by involving people, customers, partners, and suppliers in optimizing the effectiveness and efficiency demanded by our sustainability program.

Products and services are effectively promoted and marketed without making any claims that are not clearly supported by our sustainability program.

Customer relationships are managed and enhanced by informing them of our sustainability program to build and maintain an open dialogue based on openness, transparency, and trust.

Measuring Performance of the Lead and Lag Indicators

There is a reliable way to measure the performance of an organization that has been widely used around the world. It is often referred to as performance excellence. The way it scores the lead and lag indicators is presented below.

The organization carefully gathers the data and information on each of the lead indicators, including documented information derived from the system of management, internal audits, management reviews, stakeholder engagement, and benchmarking efforts. Typically, a team of assessors will be trained on how to evaluate the information and score it using the scoring matrix. Everyone on the assessment team needs to be familiar with the information on approach, deployment, assessment, and refinement (ADAR) that is presented in Chapter 16 in order to use the scoring matrix and review this data and information.[19]

If the assessors score the components of each lead indicator, the scores are added to provide a score for each lead indicator. The scores of the lead indicators can be aggregated and divided by the number of indicators to provide a single lead indicator score. The score is usually adjusted for reporting on a 1000-point basis.

Next, the lag indicators are measured against a results scoring matrix.[20] The Baldrige scores are provided in the following results categories: "Products and Processes," "Customers," "Employees, Leadership, and Governance," and "Financial and Market. The European Foundation for Quality Management (EFQM)[21] has four result categories: "People," "Customer," "Society," and "Key Performance." All the results in a category can be scored and aggregated as a score for that category. Independent KPIs without an assigned category can also be scored. Because scores have no units of measure associated with them, it is possible to aggregate all the scores and divide by the number

> ## BOX 17.1   ESSENTIAL QUESTIONS ON TRANSPARENCY, ACCOUNTABILITY, AND PERFORMANCE
>
> Why should an organization measure the contribution of the sustainability program to overall operational performance rather than simply tracking sustainability initiative progress, one KPI at a time?
>
> Why would the sustainability program choose not to be recognized as an important driver for improved performance at the facility and corporate level by restricting its measures to lag KPIs?
>
> How do the lead indicators associated with an embedded sustainability program influence and predict the lag indicators that are selected for disclosure by the organization?
>
> How does the scoring method common to the major performance frameworks provide meaningful information for disclosure while protecting sensitive information associated with some of the sustainability initiatives?

of indicators included to produce a single score for the lag indicators, usually presented on a 1000-point basis.

The two scores can then be combined (one of the performance excellence programs[22] combines them at 60% lead indicators and 40% lag indicators based on almost 30 years of experience) to give the sustainability program a single score. This can be done on a quarterly, semiannual, or annual basis. This single performance can then be tracked over time to measure the continual improvement of the organization.

Essential questions for transparency, accountability, and performance can be found in Box 17.1.

Endnotes

1. ISO, 2010.
2. ISO, 2010.
3. Baldrige, 2013.
4. ISO, 2010.
5. ISO, 2010.
6. Duckworth and Moore, 2010.
7. ISO, 2010.
8. ISO, 2010.

 9. ISO, 2010.
10. ISO, 2010.
11. ISO, 2010.
12. ISO, 2010.
13. Kaplan and Norton, 2004.
14. Kaplan and Norton, 2004.
15. ISO, 2015.
16. ISO, 2014a.
17. Baldrige, 2013.
18. EFQM, 2010.
19. McCarthy et al., 2002.
20. Baldrige, 2013.
21. EFQM, 2010.
22. Baldrige, 2013.

18

Sustainability Self-Assessment and Maturity

The maturity of an organization's system of management is determined through the use of self-assessment. *Self-assessment* is defined as a comprehensive and systematic review of the organization's activities and its sustainability performance in relation to its degree of maturity. This process is used to determine the strengths and weaknesses of the organization in terms of managing its opportunities and threats. It is important for the self-assessment to cover all the internal levels of the organization, from the leaders to individuals responsible for the operations or supporting operations. The combination of self-assessment and maturity is used to help organizations prioritize, plan, and implement improvements and innovations, as discussed in Chapter 19.

The results of self-assessment should be communicated within the engagement with both the internal and external stakeholders. Self-assessments provide information about the organization and a perspective of its future direction. These are topics that are commonly discussed within a stakeholder engagement process. The results of the monitoring and measurement should be an input to the self-assessment, but should not be thought of as a replacement of a well-designed self-assessment process. Stakeholders are more interested in the maturity of an organization's sustainability program than a report that discusses the sustainability results achieved by the organization in the past.

Evaluation of Performance

Once the measurement results are available, leaders need to assess the progress of the organization in meeting its cascaded objectives by evaluating the employee goals within the action plans prepared at the beginning of each reporting period (e.g., monthly, quarterly, or annual). There are a number of ways that this evaluation is conducted.

Internal Auditing

Internal audits are commonly used in the "check" phase of an organization's plan–do–check–act (PDCA) activity. These audits help identify problems, threats, opportunities, and nonconformities with the sustainability program that is embedded in the operations and supporting operations.[1] By focusing on the system of management, these internal audits help the organization to monitor progress in its sustainability program and the success of closing identified nonconformities from previous audits. In this manner, an internal audit acts as a means of verifying that the planned sustainability actions have been effective and how these actions are improving the ability of the organization to meet its strategic and operational objectives.[2]

Internal audits are also helpful in identifying "good practices" that can be extended to other areas of the organization's operations, products, and services. This will help the organization continually improve as a result of having its sustainability program embedded within its operating system. Internal auditing is not as effective in evaluating a sustainability program that is restricted to work on initiatives rather than embedding sustainability into the way the organization operates every day. Internal audits are an essential component of how organizations manage within their system of management.

Benchmarking

Benchmarking is a measurement and analysis methodology that an organization can use to search for "best practices" within its operations or by comparing activities with those of another organization. Organizations engage in benchmarking with the aim of improving overall performance, as well as improving their policies, strategies, operations, processes, products, and services.[3]

There are many different types of benchmarking[4]:

- Internal benchmarking comparing activities within the organization
- Generic benchmarking comparing strategies, processes, and operations with those of organizations that are not competitors
- Competitive benchmarking comparing strategies, processes, and operations performance with those of competing organizations

Generic process benchmarking can follow the model of a hotel benchmarking selected processes with a hospital. They share many activities, but do not compete with each other. Competitive benchmarking is often facilitated by a trade association or through the use of an independent analyst working on a nonattributed basis. Many smaller organizations conduct more informal benchmarks within their monitoring and measurement program by talking to other organizations at meetings or by hearing the comparisons conducted by mutual customers or groups of stakeholders. Maturity information is easier to benchmark because it does not provide specific process information that could be considered proprietary in nature.

Uncertainty Assessment

Because there are changes taking place in the internal and external operating environment, it is important to perform internal audits and benchmarks on how well the organization was able to respond to these opportunities and threats once identified. This is likely to involve the auditing of the uncertainty responses selected after identifying the significant opportunities and threats. Once the uncertainty is addressed, it would be possible to conduct audits on the risk management process that is focused on determining how effective the organization has been in meeting its strategic objectives.

Self-Assessments

A self-assessment is defined as a "cyclic, comprehensive, systematic and regular review of an organization's activities and results against a model of organizational maturity culminating in planned improvement activities."[5] The cyclic model is very similar to the PDCA cycle (Figure 18.1).

The self-assessments will be performed on each of the elements of the system of management in use by the organization:

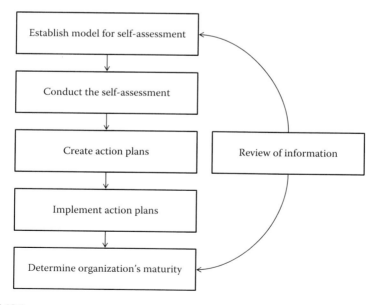

FIGURE 18.1
Self-assessment process. (Adapted from van der Wiele, T., et al., *Quality Management Journal*, 7(4), 6–22, 2000.)

- Organizational objectives, worker goals, and action plans
- Scanning of the internal and external operating environment
- Engagement with stakeholders and the social license to operate
- Organizational leadership and governance
- Uncertainty assessment and significant opportunities and threats
- Operational processes and supporting processes
- Managing the system of operations
- Scoping the monitoring and measurement process
- Monitoring and measurement
- Transparency, accountability, and performance
- Self-assessment and maturity
- Continual improvement, innovation, and learning
- Managing organizational resilience

The self-assessment is related to the internal management of the sustainability program as embedded in each of the operational areas listed above. The purpose of conducting the self-assessment is to direct, focus, and motivate various improvement activities and to seek means of innovating and furthering the organization's learning process, including its knowledge management activity, sense making, and decision making. Self-assessment is linked to the operation of the system of management and enabling the organization to seek interconnections between each of the areas in this system. There are many external reasons to perform a self-assessment (e.g., stakeholder engagement, customer retention, and managing uncertainty). However, many organizations conduct their self-assessments more focused on the internal reasons.[6]

Studies have demonstrated that self-assessments provide managers with a valuable method to coordinate and provide directions to all the activities that are part of the organization's system of operation.[7] The self-assessments help increase awareness of the importance of effective processes, efficient processes, and efficacious strategy. They are seen as a way to improve operational performance.

Overall, self-assessment appears to help organizations manage data gathering, discuss strengths and weaknesses, develop an improvement plan, and link all these activities to the means of documenting the effectiveness of the sustainability policy and strategy. These activities promote organizational learning through communication of the assessment results and getting feedback from management on the results. The process provides a means for learning the importance of cascading the objectives through the organization and using employee goals and action plans to demonstrate that the objectives are being adequately addressed.

Organizations that use self-assessment reported that the strongest improvements were[8]

- Improving the reliability of operations and support processes
- Creating a better understanding of the importance and interconnectedness of the elements in the organization's system of management
- Realizing how important it is to maintain and improve the system of management
- Being able to prevent operation losses and realize the value of the significant opportunities from the uncertainty assessment
- Realizing both the cost savings of operational efficiencies and the costs associated with not embedding sustainability into the operations

Self-assessment is a process that leaders use in organizations that is aimed at increasing the effectiveness of processes and the efficiency of processes, and creating an efficacious strategy. The self-assessment can be even more effective if combined with the determination of the maturity of the organization's sustainability program.

Creating a Maturity Matrix

Maturity matrices have been used for systems of management in the quality management field for nearly 40 years.[9] The quality maturity matrix was described as a comparison measurement tool that is used to compare different operations, keeping in mind that the purpose of the comparisons was to focus attention on the areas that were not among the top-performing operations. They are also useful for reporting the comparative results of the program over time.

A maturity matrix was presented in an early sustainable development management system standard.[10] The rationale for inclusion of the maturity matrix in this standard was to help the organization determine its position along the sustainable development path. It was noted that the maturity matrix is easy to construct and maintain by any type or size organization.

It would make sense that using a maturity matrix provides an effective means for measuring the effectiveness of a sustainability program by using the system of management provided in this book. This approach provides the means for measuring the program itself, rather than its drivers or results.

An organization with a mature embedded sustainability program performs its operations effectively and efficiently and achieves sustainability success by[11]

	Level 1	Level 2	Level 3	Level 4	Level 5
Element 1	Criteria 1 base level			→	Criteria 1 best practice
Element 2	Criteria 2 base level				Criteria 2 best practice
Element 3	Criteria 3 base level				Criteria 3 best practice

FIGURE 18.2
Maturity matrix. (Adapted from ISO, Managing for the sustained success of an organization: A quality management approach, ISO 9004, ISO, Geneva, 2009.)

- Engaging with the internal and external stakeholders
- Monitoring changes in the organization's internal and external context
- Identifying possible scenarios for improvement and innovation
- Defining and deploying policies and strategies through the leaders
- Matching its strategic objectives with operational and tactical objectives
- Constantly improving its supporting process performance with lead indicators
- Managing the processes, resources, and interactions in the suppliers, inputs, process, outputs, and customers (SIPOC) level of the value chain model

Most maturity matrices use five maturity *levels* (Figure 18.2). The first column contains the program *elements* of the system of management. The performance *criteria* are placed in each row to describe the different maturity levels and determine the strengths and weaknesses. The criteria provided for the level 5 case help the organization understand its need to improve over time. Many organizations use a criteria matrix for a couple of years and then update the criteria to make the level 5 attainment to be a bit more challenging. This supports continual improvement, innovation, and learning within the organization.

Using the Maturity Matrix

There are a number of different ways to use the maturity matrix for an organizational self-assessment. Some organizations have the leaders conduct the assessment. In other organizations, employees receive training to consistently use this self-assessment method and report the results to management.

Larger organizations with a large number of facilities might have the internal audit teams conduct the maturity assessment. If the use of the maturity matrix is independent of the organization's people, it is important that operational management and process owners take part in the evaluation to ensure that the process will capture the organization's behavior and current performance.[12]

The maturity matrix is typically used in a step-by-step process[13]:

1. Define the scope of the self-assessment by deciding what parts of the organization will be included, along with the documentation of the rationale for the selection. The focus of the assessment should also be determined and documented. Here are a few different options:

   a. A self-assessment of the key elements

   b. A self-assessment of the details of the key elements based on the specifications in place (formal or informal) of the organization's system of management

   c. A self-assessment that takes into account different levels of maturity based on a specific standard or benchmark

2. Identify the roles, responsibilities, and authorities of the people that will be involved in the self-assessment and determine when it will be conducted.

3. Determine how the self-assessment will be conducted. Who will staff the team that will be used, and will there be a facilitator assigned to provide oversight to the process?

4. Identify the maturity level for each of the organization's individual processes. This is accomplished by comparing what is observed during the assessment with what is listed in the maturity grids, and by marking the elements that the organization is already applying. The current maturity level will be the highest maturity level achieved with no preceding gaps up to that point.

5. Consolidate the results into a report. It should provide a record of progress over time and can be used to present information both internally and externally.

6. Assess the current performance of the organization's processes and operations identifying the strengths and opportunities for improvement.

An organization can be at different maturity levels for different elements. A review of the gaps can help leaders plan and prioritize the improvement or innovation activities required to move individual elements to a higher level. It is important to see the results of the maturity grid review as the "act" portion of the PDCA cycle. Each report will be used to start the cycles planning option for yet another time to drive the continual improvement.

Capability Maturity Model

Maturity grids should *not* be confused with capability maturity models. For many people, differentiating between capability maturity models and maturity grids is difficult. It is instructive to look as the key distinctions between the two.[14]

Work Orientation

Maturity grids apply to organizations and do not specify what a particular process should look like. The purpose is to identify the characteristics that any process should have to expect a desired outcome. A *capability maturity model* identifies the best practices for specific processes and evaluates the maturity of an organization in terms of how many of the practices it has implemented. These models are more of an industrial assessment.

Mode of Assessment

An assessment using a *maturity grid* is structured by using the matrix itself. Levels of maturity are allocated against key aspects of performance or key activities that are included in the matrix cells. It is easy to see how the *performance varies by different characteristics of the program*. A *capability maturity model* assessment uses questionnaires and checklists to facilitate the assessment of overall performance of an organization.

Intent

Maturity grids tend to be somewhat less complex as diagnostic and improvement tools without aspiring to provide certification. Many *capability maturity models* follow a standard format and are internationally recognized. It is easy to use them for the certification of the organization's performance.

It is possible that some business organizations may be using both of these methods.

Maturity Plots

By using the level numbers as scores, it is possible to create radar (spider) plots that are useful in communicating the results. Radar plots provide a useful way to display the maturity measurements so that they can be visually compared with each other (Figure 18.3). It is possible to have different

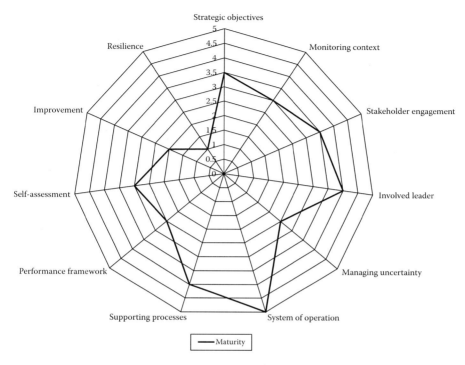

FIGURE 18.3
Radar plot of the maturity of the program elements.

plot lines for different periods of measurement or comparing the actual to the planned amounts.

In the self-assessment covered in this chapter, we are measuring the elements of the sustainability program. However, this plot could also be used for measurements of the maturity of the use of the sustainability principles (Chapter 2) or the scores of the six lead indicators (Chapter 17). Various scores can be used at any level in the organization or for the organization as a whole. This does not need to be limited to the maturity measurement.

The maturity measurement represents a much better way to express the sustainability program within the stakeholder engagement process. It is really difficult for stakeholders to go through a number of initiative results and try to figure out just how well everything is going within the company. These maturity examinations and the resultant plots should help to detect trends that conditions are improving or that something is backsliding and needs improvement. Quality management programs use these methods to detect changes in the products and services before they lead to problems in the marketplace. The same should be true with sustainability programs.

BOX 18.1　ESSENTIAL QUESTIONS ON SUSTAINABILITY SELF-ASSESSMENT AND MATURITY

Why should an organization consider using self-assessment as a means for measuring the performance of its sustainability program?

What other performance tools can be used to support the measurement of the performance of an organization's sustainability program?

Why is a maturity matrix an effective means for an organization to continually improve its sustainability program?

What are some of the uses for a maturity radar (spider) plots for communicating the status of a sustainability program within the stakeholder engagement process?

Essential questions for sustainability self-assessment and maturity can be found in Box 18.1.

Endnotes

1. ISO, 2011.
2. ISO, 2009.
3. ISO, 2009.
4. Pojasek, 2010a.
5. van der Wiele et al., 2000.
6. van der Wiele et al., 2000.
7. van der Wiele et al., 2000.
8. van der Wiele et al., 2000.
9. Crosby, 1979.
10. BS, 2006.
11. ISO, 2009.
12. ISO, 2009b.
13. ISO, 2009b.
14. Maier et al., 2012.

19

Improvement, Innovation, and Learning

Organizations need to meet their strategic objectives in an uncertain world. Improvement is essential for an organization to maintain its operational performance and react to changes in the organization's internal and external operating environments. As such, improvement is essential for an organization to exploit in order to maintain current levels of performance. It is a key part of the "act" component of the plan–do–check–act (PDCA) cycle—taking actions to continually improve.

External change is a driver for innovation. Changes in the organization's external operating environment could require innovation in order to address the interests of the stakeholders. An organization needs to identify the need for innovation and establish an effective innovation process.

Organizational learning includes both continual improvement of existing operations and significant change or innovation often leading to new objectives, products, and services. Through learning, an organization enhances, captures, and formulates the knowledge, skills, and experience of its members. This learning can be integrated into what is known as organizational wisdom, which is shared and used by the organization to support and enrich its improvement and innovation processes.

Organizations that are embedding sustainability should be able to balance and increase the satisfaction of the full range of stakeholders. These organizations need to continuously develop their capabilities to improve, innovate, and embed these activities into their products, processes, operational structures, way of operating in their context, and system of management. The foundation for effective and efficient improvement and innovation processes is learning.

Continual Improvement

An organization defines its strategic objectives to seek improvement of processes, products, and services based on information it receives from the following activities[1]:

- Monitoring, measurement, and analysis of feedback from its full range of stakeholders
- Monitoring, measurement, and analysis of its operations and supporting processes

- Internal audits and self-assessments
- Suggestions from stakeholders and partners
- Management reviews at the operational and strategic levels

It is also possible for continual improvement to be initiated through collecting data, analyzing information, setting operational and tactical objectives, and implementing corrective actions associated with any nonconformities in the system of management.

Continual improvement is a result of a set of recurring activities that are conducted to enhance the operational performance of an organization.[2] Improvement activities can range from small-step, ongoing improvement to more important breakthrough efforts. An emphasis is placed on continual improvement in an organization through the setting of worker goals at the tactical level that help meet the tactical objectives, thus beginning the feedback loop toward meeting the organization's strategic objectives.

The terms *continuous improvement* and *continual improvement* are frequently used interchangeably. The American Society for Quality is always reminding people that there is a difference.[3] *Continual improvement* is a broader term referring to general processes of improvement and encompassing many different approaches, covering different areas at different times. *Continuous improvement* is a subset of continual improvement with a more specific focus on linear, incremental improvement within an existing process. Continual improvement is not limited to quality initiatives. Improvements can be made in organizational strategy, results, and customer, employee, supplier, and stakeholder relationships. It means "getting better all the time." Improvement is a result. It can only be claimed after there has been a beneficial change in an organization's performance.

Improvement can be effected reactively (e.g., corrective action), incrementally (e.g., continual improvement), by step change (breakthrough), or proactively (e.g., innovation).[4] Improvements can be focused on a program (e.g., environmental management), in a process (e.g., resulting in a product or service), or in results. It is common for people to look for areas of underperformance or opportunities to improve that can be addressed by the practice of continual improvement. When an improvement project begins, it is possible to track, review, and audit the planning, implementation, completion, and results of these initiatives. To address these widespread improvement activities, the quality management system of operation has raised improvement to a *principle* seeking to create a quality culture that leads to[5]

- Employing a consistent organization-wide approach to continual improvement
- Making continual improvement of products, services, processes, and systems an objective for every individual in the organization (i.e., embedded in the work)

- Providing people with training in the methods and tools of continual improvement
- Ensuring that people are competent to successfully promote and complete improvement projects
- Establishing goals to guide, and measurements to track, continual improvement
- Recognizing and acknowledging the improvements that result from these activities

In practical terms, some improvements take time to achieve. Some may require allocations for a budget or careful planning for a consistent rollout of the results. Plans for improvement should consider priorities and relative benefits and should allow for the monitoring of progress.[6] Continual improvement of the organization's overall performance should be a strategic objective of the organization.[7] It is widely believed that organizations that have an ongoing focus on improvement are successful in many of the ways that such success is measured.

The leaders of the organization need to ensure that continual improvement becomes established as part of the organization's culture through a number of directed efforts by[8]

- Providing opportunities for the internal stakeholders to embed improvement into their work
- Providing the necessary resources for this to happen
- Recognizing and rewarding successes
- Continually seeking to improve the effectiveness and efficiency of the improvement process

Breakthrough projects can be conducted using a combination of project management methods and approaches for the management of change while seeking to introduce these methods into the culture of the organization.

It might help to look at the basis for determining the maturity of an organization's innovation process[9]:

- *Level 1*: Improvement activities are ad hoc and based on customer or regulatory complaints.
- *Level 2*: Basic improvement processes, based on corrective and preventive actions, are in place. The organization provides awareness for its members to help understand the concept and practice of continual improvement.
- *Level 3*: Improvement efforts can be demonstrated in most of the processes, products, and services of the organization. The focus of the improvement process is aligned with the strategy and the

strategic objectives. Recognition systems are in place for members that are generating strategically relevant improvements in their work. Continual improvement processes work at some levels of the organization and with its suppliers and partners.

- *Level 4*: Results generated from the improvement processes enhance the performance of the organization. The improvement processes are systematically reviewed. Improvement is applied to processes, products, services, organizational structures, the operating model, and the organization's system of management and supporting processes.
- *Level 5*: There is evidence of a strong relationship between improvement activities and the achievement of above-average performance for the organization determined in benchmarking activity. Improvement is embedded as a routing activity across the entire organization, as well as for its suppliers and partners. The focus is on improving the performance of the organization, including its ability to learn and change.

Innovation

Changes in the organization's internal and external operating environments may require innovation in order to address the interests of the stakeholders and continually improve its ability to meet its strategic objectives.[10] Leaders should identify these changes and initiate the impetus for innovation through the refinement of the strategic plan so that it continues its efficacious ways.

Innovation can be applied within the organization through changes in[11]

- Technology, product, or service to not only respond to the changing needs and interests of stakeholders, but also to anticipate potential changes in the organization's external operating environment and product and service life cycles
- Processes that improve cycle time, stability, and variation
- The organization itself through innovation in organizational structure and governance
- The organization's system of management that ensures the competitive advantage is maintained and new opportunities are realized when there are emerging changes in the organization's context

Planning for innovation should drive the organization to check its ability to afford the proposed activity by prioritizing the information and resources that are needed for this purpose. The resources that the organization should consider include[12]

- Creative ideas from all the engaged stakeholders
- Results from management reviews of its strategy
- Results of activities to improve its system of management and supporting processes
- Organization's performance
- Results of assessments commissioned by the leaders
- Evaluating the threats, as well as the opportunities for innovation
- Internal skills and knowledge, as well as assistance from suppliers, partners, customers, and stakeholders
- Availability of scientific or technical information
- Products and services that are approaching the end of their productive time and the need for replacement product and service offerings
- Availability of methods for innovation while considering both the opportunities and threats

The design, implementation, and management of processes for innovation can be influenced by[13]

- The urgency of the need for innovation
- Innovation objectives and their impacts on processes, products, services, and the organization's structure
- Understanding of the organization's current situation and current capabilities in relation to its innovation objectives
- The leader's commitment to innovation
- Willingness to challenge the status quo
- Availability and emergence of new technologies
- Identification of the threats associated with innovation
- Successful exchange of knowledge and expertise internally and outside the organization

The timing for the introduction of an innovation process is usually a balance between the urgency with which it is needed and the resources available for effecting change.

Innovation that is based on the organization's learning ability is essential for the long-term success of its sustainability program. The organization should use a process that is in alignment with its sustainability strategy to prioritize innovations in conjunction with its uncertainty assessment process. Innovation must be supported with the resources necessary or appropriate.

Innovation has become an imperative for any organization operating in the changing world. Since almost all organizations are dealing with frequent and significant changes, they need a defined process for developing and maintaining innovation across everything that they do every day.

It might help to look at the basis for determining the maturity of an organization's innovation process[14]:

- *Level 1*: There is limited innovation. New products and services are introduced on an ad hoc basis with no planning for innovation in place.
- *Level 2*: Innovation activities are based on data concerning the needs and interests of customers and other stakeholders.
- *Level 3*: The innovation process for new processes, products, and services is able to identify changes in the organization's external operating environment in order to initiate the planning for innovation.
- *Level 4*: Innovations are prioritized using uncertainty assessment based on the balance between their urgency, the availability of resources, and the organization's strategy. Suppliers and partners are involved in the innovation processes. The effectiveness and efficiency of the innovation processes are assessed regularly as a part of the learning process (see the next section, "Organizational Learning"). Innovation is used to improve the way the organization operates.
- *Level 5*: Innovation activities anticipate possible changes in the external operating environment. Preventive plans are developed to avoid or minimize the identified opportunities and threats associated with the innovation activities. Innovation is applied to processes, products, services, organizational structures, the operating model, and the organization's system of management, as well as it supporting processes.

Many consider the embedding of sustainability into the organization an important part of the innovation process. The new work items for the organization may include[15]

- Facilitate reflection on the consideration of sustainability interests in the innovation process
- Assess the level of embedding of these interests in the innovation process
- Identify points of vigilance, key points to facilitate the embedding of interests at each stage of the process
- Propose an evaluation system that is consistent with sustainability interests

Innovation is essential in today's world for all organizations. This globalized context implies that the implementation of partnerships and the sharing of a common vocabulary and practices have become essential to sustainable development.

Organizational Learning

The organization should encourage improvement and innovation through learning. In this way, the organization can enhance, capture, and formulate the knowledge, skills, and experience of its members, and integrate them into organizational wisdom, which can then be shared and used by the organization to foster its improvement and innovation processes.[16]

Development of the organization's learning ability depends on its ability to collect information, analyze, and gain insights from various activities in its internal and external operating environments, as well as its ability to integrate the knowledge and thinking of its members into the value system of the organization (see Chapter 3).

To be successful, organizational learning is embedded in the way an organization operates, just like sustainability. This means that learning is[17]

- Part of what members and employees do every day
- Practiced at personal, work unit, and organizational levels
- About solving problems at their source (i.e., root cause analysis)
- Focused on building and sharing knowledge and participation in sense making
- Driven by opportunities to create significant and meaningful change and to innovate

Learning is achieved through research, evaluation, and improvement cycles; ideas from employees and external stakeholders; sharing of best practices with partner organizations; and benchmarking. Learning contributes to the organization having a competitive advantage and sustainability.

Organizational learning can result in the following[18]:

- Enhanced value to customers through new and improved products and services
- Development of new organizational and business opportunities
- Development of new, improved, or disruptive processes or business models
- Reduced errors, defects, wastes, and their related costs

- Improved process responsiveness and the cycle time required for products and services
- Increases in resource productivity resulting in requirements for less resources as inputs
- Enhanced performance in fulfilling the organization's sustainability policy

Learning that integrates the capabilities of the organization and its members can be achieved by combining knowledge, thinking patterns, and behavior patterns of people with the values of the organization.[19] This involves the consideration of[20]

- The organization's mission, vision, values, and strategy
- Supporting efforts in learning and demonstrating the leaders' involvement
- Stimulations of networking, connectivity, interactivity, and sharing of knowledge both inside and outside the organization
- Maintaining systems for learning and sharing of knowledge
- Recognizing, supporting, and rewarding the improvement of people's competence, through processes for learning and sharing of knowledge
- Appreciation of creativity, supporting the diversity of opinions of the different people in the organization

Rapid access to and use of this knowledge can enhance the organization's ability to manage and maintain its sustainability program.

It might help to look at the basis for determining the maturity of an organization's innovation process[21]:

- *Level 1*: Some lessons are learned as a result of complaints and the difficulty of maintaining the social license to operate. Learning is on an individual basis without the sharing of knowledge.
- *Level 2*: Learning is generated in a reactive way from the systematic analysis of problems and other data. Processes exist for the sharing of information and knowledge.
- *Level 3*: There are planned activities, events, and forums for sharing information. A system is in place for recognizing positive results from suggestions or lessons learned. Learning is addressed in the strategy and policies.

BOX 19.1   ESSENTIAL QUESTIONS ON IMPROVEMENT, INNOVATION, AND LEARNING

Why is it important for an organization to seek continual improvement of its processes, products, and services?

What is the key driving force that creates the impetus for an organization to improve its innovation process?

What is the relationship between improvement, innovation, and learning, and why is this relationship important to the development of effective learning?

How does a maturity matrix help an organization perform a self-assessment and measure its progress on improvement, innovation, and learning?

- *Level 4*: Learning is recognized as a key process. Networking, connectivity, and interactivity are stimulated by leaders to share knowledge in the organization. Leaders support initiatives for learning and lead the process by example. The organization's learning ability integrates personal competence and organizational competence. Learning is fundamental to the improvement and innovation processes.

- *Level 5*: The culture of learning permits the taking of risks and the acceptance of failure, provided this leads to learning from the mistakes, threats, and opportunities that are identified. There are external engagements for the purpose of learning.

Essential questions for improvement, innovation, and learning can be found in Box 19.1.

Endnotes

1. ISO, 2009b.
2. Praxiom, 2014.
3. ASQ, n.d.
4. ISO, 2015.
5. ISO, 2009b.
6. ISO, 2013.

7. ISO, 2009b.
8. ISO, 2009b.
9. ISO, 2009b.
10. ISO, 2009b.
11. ISO, 2009b.
12. ISO, 2009b.
13. ISO, 2009b.
14. ISO, 2009b.
15. ISO, 2012.
16. ISO, 2009b.
17. Baldrige, 2013.
18. Baldrige, 2013.
19. ISO, 2009b.
20. ISO, 2009b.
21. ISO, 2009b.

20

Resilient Organizations

When studying sustainability, one learns about resilient ecosystems, resilient infrastructure, resilient individuals, resilient communities, and resilient value chains. There has been a growing body of knowledge on resilient organizations. Every day, the sustainability of an organization is tested in a world that constantly changes. Resilience is characterized as a strategic objective that is intended to help an organization survive and prosper.[1] A highly resilient organization is more adaptive, competitive, agile, and robust than less resilient organizations. These organizations can anticipate, prepare for, and respond and adapt to everything from minor disruptive events to acute shocks and other effects of uncertainty.[2]

Resilience is a dynamic concept without a fixed endpoint. It is a never-ending journey that operates within a risk landscape and is conditioned by people, knowledge, and technology, as well as their interactions.[3] Resilience is enhanced by integrating and coordinating the various operational programs that are commonly found in a sustainability program. Resilience is built not only within the organization, but also across its value chains and the larger web of interactions with other organizations.

Dimensions of Organizational Resilience

The study of resilient organizations suggests that this resilience is a function of three interdependent attributes[4]:

- *Leadership and culture*: The adaptive capacity of the organization
- *Networks*: The ability to leverage the internal and external relationships that have been developed
- *Change readiness*: The results of planning and direction that helped establish the ability of the organization to be change-ready

The core strategic purpose of resilience is to enable an organization to prepare itself to be able to survive and prosper. These purposes include the following[5]:

- *Competitiveness*: Being able to continue past, recover, and learn from, and even capitalize on, opportunities presented by disruptions in a way that increases value that exceeds that of the competitors who are less resilient.
- *Coherence*: Aligns strategic objectives with operational resilience measures, such as the alignment of organizational silos to become more integrated and interoperable.
- *Efficiency and effectiveness*: Creating a framework to mesh together diverse components, while allocating resources to improve overall resilience.
- *Reputation*: A coherent framework helps build trust among the various operating units and outside allies helping to manage and enhance the organization's reputation.
- *Community resilience*: Organizational resilience helps to enhance community resilience by being able to provide assurance to stakeholders (e.g., regulators, government, customers, partners, and families) of its ability to provide vital products and services to the public in times of need.

Every organization is subject to the effects of uncertainty, and resilience thinking must embrace learning (Chapter 19). Achieving favorable outcomes in the face of uncertainty requires creativity and innovation. An organization must be able to overcome strategic barriers to answer three basic questions after experiencing a disruptive event[6]:

1. What have we learned?
2. How did we learn?
3. How can we integrate what we learned to understand complex, interrelated events; engage in reflective conversations; and cultivate shared aspirations for the future?

As organizations build on the past to expand their ability to act, they are developing new ranges of possible actions that will enable them to meet threats posed by disruptive events. They are even looking for opportunities that are often masked by the threats that demand immediate action. Responding without hesitancy to unexpected events is now considered an organizational imperative.[7]

Organizational Foundations for Resilience

In order to build resilience into the operating structure of an organization, it is necessary to create a foundation that helps define the attitudes that affect organizational decisions and actions.[8] This foundation should support the further development of resilience in organizations.

Organizations need to have the ability to detect, assess, prevent, and respond to and recover from disruptive events and challenges of all types. However, organizational resilience differs significantly from the more traditional concepts of business continuity. Resilience seeks to create critical thinking, learning, and capabilities not just to bounce back (as with the practice of continuity), but also to bounce forward. This requires a combination of continuity and adaptability. The concept of adaptability includes a strong sense of societal security where preestablished relationships, learning, flexibility, and a sense of a new normality are built into the organizations so they can participate in this new focus on continuity and adaptability.[9] After all, the organization is the basic building block of society.

It would be instructive to take the framework for sustainability and see how it can be adjusted and adapted to include this continuity–adaptability perspective.

Resilience in the Structural Operating Framework

Mission Statement and Strategic Objectives

The organization needs to have a common vision and purpose (included in the mission statement) for the future (included in the sustainability program). This enables the organization to build resilience into the strategic objectives so that challenge, change, and opportunity are assessed against the vision and mission and can be acted upon accordingly.[10]

Leaders of the organization need to understand how the organization's efficacious strategy as created for the sustainability program can address the needs of resilience development. This understanding starts with knowledge of the current level of resilience.

The organization needs to promote and create universally shared expectations that strengthen its resilience. This may be reflected in one of the strategic objectives. It can be accomplished by developing and promoting cultural norms that value openness in review and evaluation of the organization's resilience and how resilience will be addressed in the feedback loop to ensure that the strategic objectives are met in an uncertain world.

Context of the Organization

For resilience to be effective, the organization must be highly informed about its internal and external operating environments. There is a new focus on influences, factors, opportunities, and threats that might influence or compromise the organization's resilience. Sometimes, this is referred to as situational awareness in the resilience literature.

In addition to the scanning performed as part of the sustainability program, the organization needs to pay attention to methods designed to identify opportunities and threats that fall within the resilience realm[11]:

- Identifying what values and purpose that it wants to protect and any threats that could impact those set of values and purpose
- Looking for emerging factors, along with the other scanning targets, to find new opportunities and threats affecting resilience
- Drawing upon sense making with the assistance of the stakeholder engagement process and the knowledge management function that is constantly examining opportunities and threats both from the context and from the literature
- Using uncertainty assessment and response to help use the opportunity response to offset the threats before they contribute to the level of uncertainty that lies between the organization and its ability to meet its strategic objectives

It is also important to scan for the kinds of events that might be expected to affect the organization. An *event* is defined as an occurrence or change of a particular set of circumstances and the effect they can have on the organization. That event could have one or more occurrences and several causes. An event could also consist of something *not* happening. Some events do not have consequences and are referred to as a "near miss," "incident," or "close call." Environmental management and health and safety management have preparedness and emergency response programs to deal with events. There are also management programs for business continuity. All these programs will be included within the organization's response to resilience.

Stakeholders and the Social License to Operate

The stakeholders are also dealing with uncertainty. They understand how it can affect their organizations. They may not understand the power of resilience to help deal with uncertainty. Sharing information on resilience within the stakeholder engagement process will truly be valued. The ability of the stakeholders to assist with lookout for change in the external operating environment and with sense making will provide the organization with their appreciation for advice on using the adaptive capacity of resilience. By being transparent on the topic of resilience within the stakeholder engagement process, the organization can help build a greater level of trust. This should help the organization secure and maintain its social license to operate from the stakeholders.

Leadership and Commitment

Leaders need to consider the impact of their decisions and the sustainability strategy on an ongoing basis. They need to create a culture in which it is essential to consider resilience within the decision-making process (see Chapter 3). This culture of trust, openness, and innovation will empower the members or employees in an organization to assume ownership of and address uncertainty as the scans of the external operating environment note new opportunities and threats. Authority and responsibility are delegated by the leaders to the individuals best able to make the decision for the organization, both under normal operations and in crisis.

The leaders are responsible to lead the engagement with all stakeholders. The engagement process needs to foster transparency that enables information to be proactively shared across internal boundaries with independent partners and other key stakeholders.

Effective governance enables the organization to exploit the results of the uncertainty assessment and response. This should direct the internal stakeholders to make decisions in accordance with the knowledge and sense making that has been increasingly involved in assembling knowledge on resilience. Effective governance also enables the organization to encourage improvement and innovation to help continually improve the knowledge system. Resilience will succeed when the governance is coherent, transparent, and "forward looking" and is embedded in a culture that is supportive of the continual enhancement of organizational resilience.

The leaders are accountable for ensuring that an appropriate level of resilience is achieved by the organization. This accountability will begin with a resilience policy that is embedded in the sustainability policy and coordinated with other important policies. This provides the leaders' commitment to all the outcomes important to the organization, including sustainability and resilience.

Managing Uncertainty and Organizational Planning

The organization needs to adapt to changing conditions as they emerge in the internal and external operating environments. The organization may choose to accept that some disruptions cannot be prevented and will still occur. It can plan, implement, test, and review a range of measures that will help the organization prepare to deal with disruption, possibly arising from unforeseen events, or effectively adapt when the established plan does not cover what is being experienced.[12]

It is important that the people responsible for uncertainty assessment have open communications with the people responsible for operations, supporting processes, and innovation, along with others that need to respond to the changes. Leadership must be involved in this effort to ensure the ability to take timely and informed actions to intercept and contain adverse events, both foreseen and unforeseen, such as to effectively respond to the uncertainty, including overwhelming crises that threaten the continued existence of the organization.[13]

Organizational System of Operating

Perhaps the most important element of resilience is the need to develop coherence in the organization's operations that provide its products and services. When an organization has coherence, all its activities and processes work well together. Even simple organizations have a large number of processes to manage. The systems of management now have a harmonized structure that makes it easy to add various operational components in a freestyle, mix-and-match manner. On the other hand, it is also easy for an organization to remove operational elements that are no longer being used. Leaders need to align operational activities within this structure to achieve coherence and build resilience. To ensure that organizational silos support resilience, the organization needs to embed sustainability and risk management activities and operational disciplines as a means to drive integration. Knowledge also needs to be actively shared across internal organizational boundaries so that opportunities and threats are addressed coherently by all parts of the organization.

Sustainability and resilience draw on a large list of different processes (see Appendix I). For many organizations, these processes are operated in silos. Leaders should find a way to create bridges between these operating silos through the use of the process approach (Chapter 12).

It is important to understand the interdependencies with other organizations, including suppliers, contractors, outsourcers, and competitors.[14] There is a greater need for interdependencies when developing resilience methods.

The organization needs to adapt its operations to changing conditions as they emerge in the internal and external operating environments. This may involve switching between preplanned responses to events and adaptive actions as necessary, and modifying its governance structures, operations, and behaviors to adjust to new conditions.[15] The planning for the operations focuses on normal situations. There must be a link between the uncertainty assessment and response efforts in the planning and emergency planning for abnormal events in the operations. This will allow the organization to respond to change in a resourceful manner as the support processes help operations adjust to the new conditions.[16]

In order to strengthen the operations of the organization, specific measures need to be implemented to address disruptive events, emergent risks, and changes in the external operating environment. This can be addressed by ensuring that the employees understand the importance of considering resilience during decision making and other instances of change management. All employees may need to take actions to prevent or reduce the likelihood of disruptive events or disruption to protect people, physical assets, financial value, reputation, and social capital.

Organizational Supporting Operations

Organizations need people who are competent in resilience and adaptability through education, training, and experience to develop and implement uncertainty assessment and business continuity plans. Employees need to be able to identify significant threats and opportunities associated with their work and to apply procedures to reduce the consequences of a disruptive event. Because many organizations will not experience major disruptions, experience can be achieved through exercise and rehearsed drills. The exercises need to be modified regularly to take into account new information in the knowledge management system. However, organizations and their members must be able and willing to adapt to change in order to become resilient. When a significant disturbance strikes, continuity plans may need to be radically adapted to reflect the new circumstances. In some cases, the business continuity plans will need to be discarded to ensure that appropriate and considered action is taken.[17]

The organization needs to create the means, incentives, and imperatives to communicate information on how resilience is a unifying factor in the organization's system of management. These messages will create a need for engaged employees and outside stakeholders who want to know more. Changes associated with resilience will lead to collaboration between those involved in these stakeholder engagements, as well as collaboration across the value chain and even among competitors.

The importance of building adaptive capacity must be included in the awareness program that is operated by the supporting operations. They need to disseminate and oversee the implementation of good practice for dealing with resilience identified from within and outside of the organization. It is useful to have an effort to share information on errors, failures, and mistakes in a transparent manner so that all the internal stakeholders are aware and learn how to avoid these issues. Many organizations share information through trade and professional associations to gain valuable lessons on how to deal with uncertainty and improve operations. This activity helps all organizations improve. Finally, the adaptive capacity can be improved by developing the internal stakeholders to get involved in the innovation process and to know that there is a need to be flexible during times of change.[18]

Resilience in Organizational Performance Management

The organization needs to include efforts to address resilience when it audits its system of operation in a way that demonstrates the efficacy of the program. The resilience measures need to be tied to the capacity of people to learn and adapt when required. This information should inform the organization's understanding of its resilience capability and may be used to prompt strategic change to further enhance how resilience will drive strategic change. As part of these efforts, the organization should verify that it is complying with legal and regulatory obligations.

Monitoring and Measurement of Performance

The organization must define the transparency and levels of accountability by which individual and collective decisions and actions on resilience are related to the norms, expectations, and obligations of the organization, its partners, and its stakeholders. There needs to be a means of measuring the performance of the resilience efforts within the framework of measuring the normal operations, products, and services.

There should be a self-assessment and maturity matrix for resilience. This would enable the leaders of the organization to determine the level of resilience that is already in place and how new efforts to enhance resilience are leading to an increase in resilience maturity. To effect this assessment, the organization should identify[19]

- What needs to be monitored and measured
- The methods for monitoring, measurement, analysis, and evaluation, as applicable, to ensure valid results
- How to provide a continuous assessment of resilience
- The thresholds at which the output from measurements will be considered acceptable
- How measurement and monitoring arrangements will work alongside, support, or integrate into existing monitoring processes
- How the results from monitoring and measurement will be analyzed and evaluated

The organization needs to understand what evidence it requires to support its assessment of resilience and ensure there is an evaluation process that is developed to support the evidence.

Improvement, Innovation, and Learning

There needs to be continual improvement of the efforts of the organization to address resilience in combination with the sustainability program.

Innovation needs to be tied to the organization's adaptive capacity in its operations. During uncertainty assessment and response, there needs to be a direct link with the people responsible for the innovation (e.g., introducing new materials, ideas, or products and services). This will enable the organization to exploit the opportunities and offset the unexpected threats that arise during these scanning or sense-making activities. This may lead to

BOX 20.1 ESSENTIAL QUESTIONS ON RESILIENT ORGANIZATIONS

How can an organization introduce resilience into its strategic objectives in order to help it cope with uncertainty and unanticipated disruptions over the long term?

How can the organization build the capacity to adapt to changing conditions as they emerge?

Why is it important for the organization to bring coherence to its system of operations and supporting processes?

Why should an organization embed resilience into its system of operations rather than create a new program to address disruptions to its operations?

new or improved products and services that fit the new conditions brought about by long-term changes in the operating environment. In this way, innovation can serve the organization by identifying new and better solutions and addressing the shifting needs of the organization arising from its ever-changing external operating environment.[20]

Learning is very important in the practice of resilience. An organization should constantly be improving its knowledge and sense making to incorporate resilience in decision making.

Essential questions for resilient organizations can be found in Box 20.1.

Endnotes

1. BS, 2014.
2. BS, 2014.
3. Serrat, 2013.
4. Resilient Organizations, 2012.
5. BS, 2014.
6. Serrat, 2013.
7. Serrat, 2013.
8. BS, 2014.
9. BS, 2014.
10. BS, 2014.
11. BS, 2014.
12. BS, 2014.
13. BS, 2014.

14. BS, 2014.
15. BS, 2014.
16. BS, 2014.
17. McAsian, 2010.
18. BS, 2014.
19. BS, 2014.
20. BS, 2014.

References

AS (Standards of Australia). (2003). Good governance principles. AS 8000. Sydney: SAI Global Press.

AS (Standards of Australia). (2003a). Organizational code of conduct. AS 8002. Sydney: SAI Global Press.

AS (Standards of Australia). (2003b). Corporate social responsibility. AS 8003. Sydney: SAI Global Press.

AS/NZS (Standards of Australia/Standards of New Zealand). (2000). Environmental risk management. HB 203. Sydney: SAI Global Press.

AS/NZS (Standards of Australia/Standards of New Zealand). (2013). Risk management guidelines, companion to AS/NZS ISO 31000:2009. HB 436. Sydney: SAI Global Press.

ASQ (American Society for Quality). (n.d.). Basic concepts: Continuous improvement. Retrieved March 23, 2016, from http://www.asq.org/learn-about-quality/continuous-improvement/overview/overview.html.

Baldrige (Baldrige Performance Excellence Program). (2013). Baldrige performance excellence criteria. Gaithersburg, MD: National Institute for Science and Technology.

Bandor, M. (2007). Process and procedure definition: A primer. Pittsburgh, PA: Carnegie Mellon University. Retrieved February 26, 2016, from http://www.slideshare.net/nshahafiza/processes-and-procedures.

Barnat, R. (2014). Strategic management: Formulation and implementation. Retrieved August 3, 2015, from http://www.strategy-formulation.24xls.com.

Bennett, N., and Lemoine, G.J. (2014). What VUCA really means to you. *Harvard Business Review*, 92 (1/2) 27.

Bradley, C., Dawson, A., and Montard, A. (2013). Mastering the building blocks of strategy. *McKinsey Quarterly*, October. Retrieved February 8, 2016, from http://www.mckinsey.com.

Brown, M.G. (1996). *Keeping Score: Using the Right Metrics to Drive World-Class Performance*. New York: Quality Resources.

BS (British Standards). (2006). Guidance for managing sustainable development. BS 8900. London: British Standards Institute.

BS (British Standards). (2014). Guidance on organizational resilience. BS 65000. London: British Standards Institute.

Choo, W.C. (2001). Environmental scanning as information seeking and organizational learning. *Information Research*, 7 (1). Retrieved June 29, 2015, from http://www.informationr.net/ir/7-1/paper112.html.

Choo, W.C. (2006). *The Knowing Organization: How Organizations Use Information to Construct Meaning, Create Knowledge and Make Decisions*. 2nd ed. New York: Oxford University Press.

Community Tool Box. (n.d.). Developing an action plan. Retrieved August 3, 2015, from http://ctb.ku.edu/en/table-of-contents/structure/strategic-planning/develop-action-plans/main.

Conlon, G. (n.d.). *Capitalizing on Voice of the Customer*. Stamford, Connecticut: Peppers & Rogers Group. Retrieved from http://allegiance.com/documents/voc_ebook.pdf.

Crosby, P. (1979). *Quality Is Free: The Art of Making Quality Certain*. New York: Mentor Books.

Daft, R.L. (2013). *Organization Theory & Design*. Mason, OH: South-Western Learning, Centage Learning.

Davis, J. (n.d.). The importance and value of organizational goal setting. Retrieved June 22, 2015, from http://www.flexstudy.com/catalog/schpdf.cfm?coursenum=95086.

Dorner, K., and Edelman, D. (2015). What 'digital' really means. Retrieved August 8, 2015, from http://www.mckinsey.com/insights/high_tech_telecoms_internet/what_digital_really_means.

Duckworth, H., and Moore, R. (2010). *Social Responsibility: Failure Mode Effects and Analysis*. Boca Raton, FL: CRC Press.

Edmond, M.B., and Monroe, D.J. (2010). Empowering the workforce to tackle the useful many processes. In Juran, J.M., and DeFeo, J.A. (Eds.), *Juran's Quality Handbook*, 874–866. 6th ed. New York: McGraw-Hill.

EFQM (European Foundation for Quality Management). (2003). EFQM framework for corporate social responsibility. Brussels: EFQM.

EFQM (European Foundation for Quality Management). (2010). EFQM excellence model. Brussels: EFQM.

FMR (Free Management Ebooks). (2013). *PESTLE Analysis: Strategy Skills*. Retrieved January 31, 2016, from http://www.free-management-ebooks.com/dldebk/dlst-pestle.htm.

G31000. (n.d.). The global platform for ISO 31000. Retrieved March 9, 2016, from http://g31000.org/.

Garwood, W.R., and Hallen, G.L. (2010). *Human Resources and Quality*. In Juran, J.M., and DeFao, J.M. (Eds.), *Juran's Quality Handbook*. 6th ed. New York: McGraw-Hill.

GRI IFC (Global Reporting Initiative and International Finance Corporation). (2010). Getting more value out of sustainability reporting. Retrieved March 15, 2015, from https://www.globalreporting.org/resourcelibrary/Connecting-IFCs-Sustainability-Performance-Standards-GRI-Reporting-Framework.pdf.

Hillson, D. (2001). Effective strategies for exploiting opportunities. Presented at Proceedings of the Project Management Institute Annual Seminar and Symposium, Nashville, TN. Retrieved February 25, 2016, from http://www.risk-doctor.com/pdf-files/opp1101.pdf.

Hopkin, P. (2012). *Fundamentals of Risk Management*. 2nd ed. Philadelphia: Kogan Page.

Hubbard, D.W. (2010). *How to Measure Anything: Finding the Value of "Intangibles" in Business*. 2nd ed. Hoboken, NJ: John Wiley & Sons.

IFAC (International Federation of Accountants). (2007). Defining and developing an effective code of conduct for organizations. New York: IFAC. Retrieved February 8, 2016, from http://www.ifa.org.uk/files/PAIB%20code-of-conduct.pdf.

ISO (International Organization for Standardization). (2008). Guidance on the concept and use of the process approach for management systems. ISO/TC 176/SC2/N 544R3. Geneva: ISO. Retrieved June 22, 2015, from http://www.ios.org/iso/04_concept_and-use-of-the-process-approach_for_management_systems.pdf.

ISO (International Organization for Standardization). (2009). Risk management—Principles and guidelines. ISO 31000. Geneva: ISO.

ISO (International Organization for Standardization). (2009a). Risk management—Risk assessment techniques. ISO 31010. Geneva: ISO.

ISO (International Organization for Standardization). (2009b). Managing for the sustained success of an organization: A quality management approach. ISO 9004. Geneva: ISO.

ISO (International Organization for Standardization). (2010). Social responsibility guidance. ISO 26000. Geneva: ISO.

ISO (International Organization for Standardization). (2011). Guidelines for auditing management systems. ISO 19011. Geneva: ISO.

ISO (International Organization for Standardization). (2012). Innovation process: Interaction, tools and methods. ISO TS/P233. Geneva: ISO.

ISO (International Organization for Standardization). (2013). Risk management—Guidance for the implementation of ISO 31000. ISO/TR 31004. Geneva: ISO.

ISO (International Organization for Standardization). (2014). Quality management systems—Fundamentals and vocabulary. ISO/DIS 9000. Geneva: ISO.

ISO (International Organization for Standardization). (2014a). Environmental management systems—Requirements with guidance for use. ISO FDIS 14001. Geneva: ISO.

ISO (International Organization for Standardization). (2014b). Compliance management systems—Guidance. ISO 19600. Geneva: ISO.

ISO (International Organization for Standardization). (2014c). Asset management—Management systems—Requirements. ISO 55001. Geneva: ISO.

ISO (International Organization for Standardization). (2014d). Guidelines for addressing sustainability in standards. Guide 82. Geneva: ISO.

ISO (International Organization for Standardization). (2015). Quality management systems—Requirements. ISO 9001. Geneva: ISO.

Kaplan, R., and Norton, D. (2004). *Strategy Maps: Converting Intangible Assets into Tangible Outcomes*. Boston: Harvard Business School Press.

Kiptoo, J.K., and Mwirigi, G.M. (2014). Factors that influence effective strategic planning process in organizations. *Journal of Business Management*, 16 (6) 188–195. Retrieved February 8, 2016, from http://iosrjournals.org/iosr-jbm/papers/Vol16-issue6/Version-2/R01662188195.pdf.

Knight, K. (2010). Risk management—A journey not a destination. Retrieved February 21, 2016, from http://en.mgubs.ru/images/Image/A%20Journey%20Not%20A%20Destination.pdf.pdf.

Maier, A., Moultric, J., and Clarkson, P. (2012). Assessing organizational capabilities: Reviewing and guiding the development of maturity grids. *IEEE Transactions on Engineering Management*, 59 (1) 138–159. Retrieved March 26, 2016, from http://www.iff.ac.at/oe/full_papers/Maier%20Anja_Moultrie_Clarkson.pdf.

McAsian, A. (2010). Organisational resilience: Understanding the concepts and its application. White paper. Adelaide, Australia: Torrens Resilience Institute. Retrieved March 29, 2016, from https://www.flinders.edu.au/centres-files/TRI/pdfs/organisational%20resilience.pdf.

McCarthy, G., Greatbanks, R., and Yang, J. (2002). Guidelines for assessing organizational performance against the EFQM model of excellence using the radar logic. Working Paper Series. Manchester, UK: Manchester School of Management.

Retrieved March 25, 2016, from https://php.portals.mbs.ac.uk/Portals/49/docs/jyang/McCarthyYangGreatbanks_MSM_Guidelines_for_Selfassessment.pdf.

McChesney, C., Covey, S., and Huling, J. (2012). *The 4 Disciplines of Execution*. London: Simon & Schuster.

NORC. (n.d.). Guidelines for writing a NORC program mission statement. Retrieved June 22, 2015, from http://www.norcblueprint.org/uploads/File/Mission_Statement_Guidelines.pdf.

OECD (Organisation for Economic Cooperation and Development). (2004). *OECD Principles of Corporate Governance*. Paris: OECD Publications Service.

Pojasek, R.B. (1997). Implementing your pollution prevention alternatives. *Pollution Prevention Review*, 7 (2) 83–88.

Pojasek, R.B. (2005). Understanding processes with hierarchical process mapping. *Environmental Quality Management*, 15 (2) 79–86.

Pojasek, R.B. (2005a). *Making the Business Case for EHS*. Old Saybrook, CT: Business and Legal Reports.

Pojasek, R.B. (2006). Putting the hierarchical process maps to work. *Environmental Quality Management*, 15 (3) 73–80.

Pojasek, R.B. (2007). A framework for business sustainability. *Environmental Quality Management*, 17 (2) 81–88.

Pojasek, R.B. (2010). Sustainability: The three responsibilities. *Environmental Quality Management*, 19 (3) 87–94.

Pojasek, R.B. (2010a). Benchmarking to sustainability in four steps. *Environmental Quality Management*, 20 (2) 87–94.

Pojasek, R.B. (2012). Understanding sustainability: An organizational perspective. *Environmental Quality Management*, 21 (3) 93–100.

Pojasek, R.B., and Hollist, J.T. (2011). Improving sustainability results with performance frameworks. *Environmental Quality Management*, 20 (3) 81–96.

Porritt, J. (2006). *Capitalism as if the World Matters*. Sterling, VA: Earthscan.

Porter, M.E. (1985). *Competitive Advantage: Creating and Sustaining Superior Performance*. New York: Free Press.

Praxiom. (2014). Plain English definition ISO 9000:2015. Calgary, AB: Praxiom Reseach Group Ltd. Retrieved January 27, 2016, from http://www.praxiom.com/iso-definitions.htm.

Praxiom. (2015). ISO's process approach translated into plain English. Calgary, AB: Praxiom Research Group Ltd., Retrieved February 25, 2016, from http://www.praxiom.com/process-approach.htm.

Praxiom. (2015a). ISO FDIS 9001 2015 translated into plain English. Calgary, AB: Praxiom Reseach Group Ltd. Retrieved February 25, 2016, from http://www.praxiom.com/iso-9001.htm.

Project Management Institute. (2013). *A Guide to the Project Management Body of Knowledge*. 5th ed. Newtown Square, PA: Project Management Institute.

Purdy, G. (2014). Designing good procedures. Technical Note. Cammeray, Australia: Broadleaf Capital International. Retrieved February 26, 2016, from http://broadleaf.com.au/resource-material/designing-good-procedures/.

Resilient Organizations. (2010). Resilient organizations. Retrieved March 29, 2016, from http://www.resorgs.org.nz/Content/what-is-resilience.html.

SAI Global. (2005). Introduction to risk management. Course. Sydney: SAI Global. Current course retrieved March 9, 2016, from http://training.saiglobal.com/tis/promotion.aspx?id=a0c2000000058FuAAI.

SAI Global. (2007). *Australian Business Excellence Framework*. Sydney: SAI Global.

Serrat, O. (2008). Culture theory. Knowledge Solutions. Manila: Asian Development Bank. Retrieved from http://www.adb.org/sites/default/files/publication/27578/culture-theory.pdf.

Serrat, O. (2009). Asking effective questions. Knowledge Solutions. Manila: Asian Development Bank. Retrieved June 29, 2015, from http://adb.org/sites/default/files/pub/2009/asking-effective-questions.pdf.

Serrat, O. (2009a). Learning from evaluation. Knowledge Solutions. Manila: Asian Development Bank. Retrieved June 29, 2015, from http://adb.org/sites/default/files/pub/2009/learning-from-evaluation.pdf.

Serrat, O. (2009b). Understanding complexity. Knowledge Solutions. Manila: Asian Development Bank. Retrieved June 29, 2015, from http://digitalcommons.ilr.cornell.edu/intl/188/.

Serrat, O. (2009c). From strategy to practice. Knowledge Solutions. Manila: Asian Development Bank. Retrieved June 29, 2015, from http://digitalcommons.ilr.cornell.edu/cgi/viewcontent.cgi?article=1154&context=intl.

Serrat, O. (2009d). Building a learning organization. Knowledge Solutions. Manila: Asian Development Bank. Retrieved June 29, 2015, from http://www.adb.org/sites/default/files/publication/27563/building-learning-organization.pdf.

Serrat, O. (2009e). A primer on organizational culture. Knowledge Solutions. Manila: Asian Development Bank. Retrieved June 29, 2015, from http://www.adb.org/sites/default/files/publication/27623/primer-organizational-culture.pdf.

Serrat, O. (2010). Bridging organizational silos. Knowledge Solutions. Manila: Asian Development Bank. Retrieved June 29, 2015, from http://www.adb.org/sites/default/files/publication/27562/bridging-organizational-silos.pdf.

Serrat, O. (2010a). A primer on corporate values. Knowledge Solutions. Manila: Asian Development Bank. Retrieved June 29, 2015, from http://www.adb.org/sites/default/files/publication/27622/primer-corporate-values.pdf.

Serrat, O. (2011). Critical thinking. Knowledge Solutions. Manila: Asian Development Bank. Retrieved June 29, 2015, from http://digitalcommons.ilr.cornell.edu/intl/103/.

Serrat, O. (2011a). A primer on corporate governance. Knowledge Solutions. Manila: Asian Development Bank. Retrieved June 29, 2015, from http://digitalcommons.ilr.cornell.edu/cgi/viewcontent.cgi?article=1106&context=intl.

Serrat, O. (2012). On organizational configurations. Knowledge Solutions. Manila: Asian Development Bank. Retrieved June 29, 2015, from http://www.adb.org/sites/default/files/publication/29644/organizational-configurations.pdf.

Serrat, O. (2012a). On decision making. Knowledge Solutions. Manila: Asian Development Bank. Retrieved June 29, 2015, from http://www.adb.org/sites/default/files/publication/30024/decision-making.pdf.

Serrat, O. (2013). On resilient organizations. Knowledge Solutions. Manila: Asian Development Bank. Retrieved June 29, 2015, from http://digitalcommons.ilr.cornell.edu/cgi/viewcontent.cgi?article=1275&context=intl.

Step Change in Safety. (n.d.). *Leading Performance Indictors: Guidance for Effective Use*. Aberdeen, UK: Step Change in Safety. Retrieved February 10, 2016, from https://www.stepchangeinsafety.net/node/2667.

Talbot, J. (2011). The external context—What's outside the door? Blog. Retrieved June 29, 2015, from http://31000risk.blogspot.com/2011/04/532-external-context-whats-outside-door.html.

Talbot, J. (2011a). Internal context. Blog. Retrieved June 29, 2015, from http://31000risk. blogspot.com/2011/05/533-internal-context-html.

Thompson, I., and Boutilier, R. (n.d.). The social license to operate. Website. Retrieved August 8, 2015, from http://sociallicense.com.

Thor, C.G. (1994). *The Measures of Success: Creating a High Performing Organization.* Essex Junction, VT: Oliver Wight Publications.

Tophoff, V. (2015). From bolt-on to built-in: Managing risk as an integral part of managing an organization. New York: International Federation of Accounting. Retrieved February 29, 2016, from https://www.ifac.org/publications-resources/bolt-built.

van der Wiele, T., Brown, A., Millen, R., and Whelan, D. (2000). Improvement in organizational performance and self-assessment practices by selected American firms. *Quality Management Journal*, 7 (4) 6–22.

Appendix I

System of Management for Sustainability

The structure and content for a sustainability program can be created in the same plan–do–check–act (PDCA) structure as the harmonized high-level structure of the standards created by the International Organization for Standardization (ISO). This is great news since many organizations already use this information. ISO has been dedicated to making their information usable in any organization, regardless of its size. These latest improvements enable organizations to mix and match elements from different standards to create a program that meets their sustainability needs. In Chapter 20, the concept of "coherence" is presented as a means of helping these seemingly disparate elements work together to create the support for the development and implementation of a sustainability program. Many of these standards can be found in the reference sections.

ISO has some great publications for small organizations to use as a guide to creating a system of management that can serve as the home for a sustainability program. Most of its management system standards can be viewed at no cost on its website. While the guidance documents are for sale through ISO and its member standard-setting organizations in more than 150 different countries, there are books available to walk you through that information. The purpose of this book is to point out which standards are potentially applicable to coherently state a sustainability program that can be embedded in an organization's daily tasks. No effort was made to provide the necessary details that ISO and other standard-setting organizations have already compiled. Refer to the original information to put together a sustainability program that works best for your organization.

System of Management: ISO Annex SL and Other Standards

ISO Annex SL Processes	Other Standards
4 Context of the Organization	
4.1 Understanding the Organization and Its Context	Organization's Operating Environment (ISO 9004:2009)
4.2 Understanding the Interests of the Stakeholders	Stakeholder Identification and Engagement (ISO 26000:2010)
4.3 Determining the Scope of the Integrated System of Management	Managing for the Sustained Success of an Organization (ISO 9004:2009)
	Relationship of an Organization's Characteristics to Social Responsibility (ISO 26000:2010)
4.4 Integrated Operational System of Management	Quality Management System and Its Processes (ISO FDIS 9001:2015)
	Compliance Management System and Principles of Good Governance (ISO 19600:2014)
	Practices for Integrating Social Responsibility throughout an Organization (ISO 26000:2010)
	Identification, Analysis, and Evaluation of Compliance Risks (ISO 19600:2014)
5 Governance and Leadership	
5.1 Leadership and Commitment	Organizational Governance (ISO 26000:2010)
	Customer Focus (ISO FDIS 9001:2015)
5.2 Policy	Strategy and Policy (ISO 9004:2009)
	Developing and Communicating the Policy (ISO FDIS 9001:2015)
	Quality, Environmental, Compliance Policy
5.3 Organization Roles, Responsibilities, and Authorities	Assigning Responsibility for Compliance in the Organization (ISO 19600:2014)
	Governing Body and Top Management Role and Responsibility (ISO 19600:2014)
	Designation of Compliance Function (ISO 19600:2014)
	Management Responsibilities (ISO 19600:2014)
	Employee Responsibilities (ISO 19600:2014)
6 Planning	
6.1 Actions to Address Threats and Opportunities	Significant Aspects (ISO FDIS 14001:2015)
	Planning for Social Responsibility Core Subjects
	• Human Rights
	• Labor Practices
	• Environment

System of Management: ISO Annex SL and Other Standards

ISO Annex SL Processes	Other Standards
	• Fair Operating Practices
	• Consumer Issues
	• Community Involvement and Development
6.2 Operational Objectives and Planning to Achieve Them	Environmental Aspects (ISO FDIS 14001:2015)
	Compliance Obligations (ISO FDIS 14001:2015)
	Identification and Maintenance of Compliance Obligations (ISO 19600:2014)
	Planning for Changes (ISO FDIS 9001:2015)
7 Supporting Processes	
7.1 Resources	Resource Management (ISO 9004:2009)
	Natural Resources (ISO 9004:2009)
	Purchasing
	People (ISO FDIS 9001:2015)
	Infrastructure (IS0 9004:2009)
	Work Environment (ISO 9004:2009)
	Knowledge, Information, Technology (ISO 9004:2009)
	Quality Management Principles (ISO FDIS 9001:2015)
	Principles of Social Responsibility (ISO 2600:2010)
	Monitoring and Measuring Resources (ISO FDIS 9001:2015)
7.2 Competence	Management of People (ISO 9004:2009)
	Competence of People (ISO 9004:2009)
	Involvement and Motivation of People (ISO 9004:2009)
7.3 Awareness	Behaviors (ISO 19600)
	Awareness (ISO 9001)
7.4 Communication—Internal and External	Communication on Social Responsibility (ISO 26000:2010)
7.5 Documented Information	General (ISO 9001)
	Creating and Updating (ISO 9001)
	Control of Documented Information (ISO 9001)
8 Operation	
8.1 Operational Planning	Process Management (ISO 9004:2009)
8.2 Operations Control	Emergency Preparedness and Response (ISO FDIS 14001:2015)
	Requirements for Products and Services (ISO FDIS 9001:2015)

System of Management: ISO Annex SL and Other Standards

ISO Annex SL Processes	Other Standards
	Design and Development of Products and Services (ISO FDIS 9001:2015)
	Control of Externally Provided Processes, Products and Services (ISO FDIS 9001:2015)
	Production and Service Provision (ISO FDIS 9001:2015)
	Release of Products and Services (ISO FDIS 9001:2015)
	Control of Nonconforming Outputs (ISO FDIS 9001:2015)
	Suppliers and Partners (ISO 9004:2009)
	Outsourced Processes (ISO 19600:2014)
9 Performance Evaluation	
9.1 Monitoring, Measurement, Analysis and Evaluation	Key Performance Indicators (ISO 9004:2009)
	Evaluation of Compliance (ISO FDIS 14001:2015)
9.2 Internal Audit	Self-Assessment (ISO 9004:2009)
	Benchmarking (ISO 9004:2009)
9.3 Management Review	
10 Improvement	
10.1 Nonconformity and Corrective Action	Escalation (ISO 19600:2014)
10.2 Continual Improvement	Reviewing and Improving an Organization's Actions and Practices Related to Social Responsibility (ISO 26000:2010)
	Innovation (ISO 9004:2009)
	Learning (ISO 9004:2009)

Appendix II: Olive Hotel Case—Part 1

This appendix presents the "plan" and "do" elements of a virtual case developed by Cherie Mohr in a course that was taught using the method presented in this book. The information presented was written before the final draft of this book was completed; however, the basic organizational sustainability framework presented in the book is followed. Because this book is subtitled "Practical Step-by-Step Guide," this appendix can be used by the reader to develop some skills using the information presented in Chapters 7 through 13, as supplemented by the information in the foundation section, Chapters 1 through 6.

This practice is presented in two steps. The first step involves editing the information presented using the concepts outlined in the book. Once this has been completed, the reader should determine if each of the "essential questions" at the end of Chapters 7 through 13 are sufficiently addressed in the Olive Hotel case. Readers can also use Chapter 14 in the practice since it presents the need for interconnections between the materials presented in separate chapters in that section.

While it is important to have knowledge about the elements of a sustainability program, it is very important to develop the skills for using this knowledge for an organization that you are working with. Once you have completed this exercise for a case that you select, you should be able to develop the competence to apply the book's content to any organization operating at the community level. Unfortunately, there are very few cases available at this time that use this method for planning, implementing, and maintaining a fully embedded organizational sustainability program. More information may be found in the standards and reports described in Appendix I.

Introduction

The following is an example of a virtual case for creating a sustainable organization—a virtual hotel—with the context of a neighborhood district in Cambridge, Massachusetts. The information in this Appendix covers the structure topics (Chapters 7 through 13). The monitoring and measurement topics (Chapters 15 through 18). For each of the 16 topics, the Olive Hotel is evaluated and recommendations are made to assist the organization to become more sustainable. The Olive Hotel's name and characteristics are not based on any actual facility. The presentation is used to provide a step-by-step guide to applying the concepts in this book to actual organizations. Information presented in this case was derived from many students taking the "Fundamentals of Organizational Sustainability" course at the Harvard Extension School. These concepts can be applied to any organization of any size.

For the purposes of this case, the Olive Hotel was considered to be a property in the Accor Hotel Group, an international hotel with operations in 92 countries with more than 3700 hotels and nearly 45 years of experience.[1] Accor Group was selected since it is committed to sustainability through awareness and environmental improvement initiatives.[2] Accor Group did not participate in the development of the case. Information was used from its public website. We did not check the accuracy of this information. The purpose of using a corporate parent was solely to demonstrate how it would interact with its facilities within their community surroundings. The Olive Hotel shares the values of its parent organization (i.e., through its internal context) and seeks to develop its own unique sustainability management program in the context of the Cambridge, Massachusetts, community.

For the purposes of this case, the Olive Hotel occupies an old Victorian-style fire station, built in 1895 in the Queen Anne style. A renovation and seven-story addition, completed in 2000, converted the firehouse into the 73-room Olive Hotel. The hotel has a staff of 32 people who manage, maintain, and secure its guest rooms; reception area; three parking spaces; three meeting rooms; a restaurant that offers breakfast, lunch, and dinner; and a hotel laundry facility.[3]

The Cambridgeport neighborhood[4] is located just north of the Charles River and the city of Boston. It covers 0.53 square miles, comprising 8.3% of the area of the city of Cambridge. This neighborhood is bounded on the north by Massachusetts Avenue and Central Cambridge, on the east by the railroad tracks running parallel to the Massachusetts Institute of Technology (MIT) campus, and on the west by River Street. Cambridgeport is a relatively affluent community with a median family income that is nearly 50% higher than the national average. Due to limited street parking and the availability of public transportation, the operation of vehicles by Cambridgeport residents is less than the national average. Only 66% of Cambridgeport's approximately 5000 households owned a vehicle in 2010,[5] compared with 91% of households nationwide in 2009.[6] Parking at the hotel for visitors, including hotel patrons, is extremely limited. Valet parking is made available using some public parking garages in the area.

The commonwealth of Massachusetts imports its power from other states and Canada. In 2009, the energy source profile for Massachusetts was 57% coal and oil, 26% natural gas, and 17% nuclear, solar, and other renewables.[7] While there is a trend lowering the contribution of coal and oil, there has been an increase in greenhouse gas emissions because of the rapid growth in the city of Cambridge.[8] Reduction of total greenhouse gas emissions is a very high priority for the city of Cambridge.

In a few cases, from another "virtual" case used in the course—SMART High School—will be used in situations to further clarify the examples provided for the Olive Hotel (Box II.1).

Organization's Objectives and Goals (Chapter 7)

The strategic objectives for the Olive Hotel come from the mission statement of the Accor Group. These objectives are cascaded down through the corporation. Each objective of the Olive Hotel must have a link to the strategic objectives of the parent organization. Goals are established at the worker level at the Olive Hotel. These goals and the associated action plans help the Olive Hotel meet its overarching goals. Sustainability can be addressed in the goals as long as they contribute to meeting the objectives.

Mission Statement for the Olive Hotel

Using the following guidelines, the Olive Hotel shall create a mission statement that helps make a connection to the parent organization's mission statement. This organizational mission statement takes into account the unique internal and external context of the Olive Hotel. Having a mission statement and a set of overarching objectives helps the Olive Hotel align itself through its internal context with the Accor Group and through its external context with the Cambridgeport district in the city of Cambridge. In preparing the mission statement, the Olive Hotel should focus on a 5- to 10-year time frame. This organizational mission statement must provide a summary of the hotel's principles and culture (see Chapter 2) since this is important to the stakeholders and the securing of the hotel's social license to operate. The following guidelines should be addressed in this process:

- In cooperation with the stakeholders, management shall identify the principles and culture associated with the Olive Hotel and emphasize the values it shares with the Accor Group. For example, the hotel's mission "to provide outstanding accommodations and maximize economic opportunities without compromising its sustainability commitment" aligns well with Accor's mission of providing the best choice of services for its guests.[9]
- Describe the Olive Hotel's reason for existing. Consider how, as the only hotel of its size and character in Cambridgeport,[10] it might shape its new mission more specifically than its current stated mission, "to be of value as a community member and integrate itself into its social and ecologic environment."
- Looking forward, describe how the hotel will embed its sustainability objectives and goals into its daily operations, with special attention to the interests of its stakeholders. Consider how Accor's "five core values" can be adapted to create core values that apply to everyday operations.[11] In order to bestow a social license to operate, the community wants the hotel to be a good neighbor and a positive

contributor (i.e., socially engaged), add value to the community (i.e., economically successful), and use natural resources wisely (i.e., environmentally responsible).

- Describe how the Olive Hotel will set responsible operating objectives that will enable it to transparently manage its responsibilities for environmental stewardship, social well-being, and economic prosperity over the long term while being held accountable to its stakeholders.
- Finally, the Olive Hotel should seek to align its mission with the mission of its external context (i.e., the city of Cambridge), regarding economic and social sustainability,[12] in order to meet its objectives and thrive in its community.

The following is an example of a mission statement for the Olive Hotel:

> The mission of the Olive Hotel is to provide affordable and healthy accommodations in a comfortable and friendly setting. The Olive Hotel is committed to embedding the three responsibilities of sustainability (social well-being, economic prosperity, and environmental stewardship) into all its operations and policies. The hotel is committed to active engagement with neighbors and stakeholders while working to create trust and shared value with its host community and society at large.

Establishing Strategic Objectives for the Olive Hotel

The purpose of an organization is to achieve its objectives. These objectives arise from and give effect to the organization's mission. Based on the mission statement, the Olive Hotel should establish three to five concise strategic objectives or statements of intent, which provide a means to accomplish each aspect of the stated mission within the same time frame of 5–10 years. Operational objectives, goals, and processes will be developed from these strategic objectives. This statement should include the following considerations:

- The strategic objectives should be continuous. In other words, there would never be a time when the Olive Hotel would not want to seek to fulfill these three or four objectives.
- Taken as a whole, the strategic objectives should define the outcomes sought through the hotel's operations and explain what the hotel is trying to accomplish.
- The strategic objectives should seek to align with the desired outcome of creating value for the Olive Hotel.
- The strategic objectives should consider adopting the deeply embedded and socially, economically, and environmentally responsible sustainability objectives of the Accor Group.

BOX II.1   MISSION STATEMENT OF SMART HIGH SCHOOL

The mission of SMART High School is to provide a healthy and safe environment where curiosity, integrity, and respect guide the endeavors of every student, teacher, and staff member. SMART High School is committed to ensuring its future success by embedding the three responsibilities of sustainability, *social well-being*, *economic prosperity*, and *environmental stewardship*, into its decision making, operations, and curriculum. Sustainability is embedded into the curriculum through *integrated learning* among all disciplines through engagement between and among our students, teachers, staff, partners, mentors, host community, and the natural environment.

OBJECTIVES OF SMART HIGH SCHOOL

1. Instill curiosity, integrity, and respect into every interaction between and among students, faculty, and staff by teaching each individual how to lead by example.
2. Encourage academic excellence by teaching every student *how* to learn and integrate knowledge, rather than merely storing information.
3. Design an *integrated learning* experience for students in which teachers, staff, and mentors work with each other to create linkages in knowledge by engaging students in open discussions about these linkages.
4. Use the SMART sustainability plan to integrate sustainability into the curriculum through engagement between and among our students, teachers, staff, partners, mentors, host community, and the natural environment.

- The strategic objectives should consider Accor's sustainable development objectives[13] associated with its Planet 21 Program.

A set of strategic objectives and accompanying mission statement are illustrated for a high school in Box II.1.

Operational Objectives for the Hotel's Laundry Operation

The laundry operations of the hotel represent a level of operations the objectives need to reach, while recognizing the interests of stakeholders and the opportunities and threats posed by the internal and external context. Using the following procedure, the hotel can create two or three operational objectives for the laundry operation:

- Define the health and safety objectives for creating not only a suitable work environment,[14] but also a legally compliant work environment.[15]
- Define environmental efficiency objectives by identifying ways to save water,[16] detergents,[17] and energy[18] and conserve materials.[19]

It is important to remember that the operational objectives support the organization's strategic objectives. When there is a parent organization, the organization's strategic objectives must be aligned with the corporate objectives. The goals and action plans are used at each level in an organization to help demonstrate progress in meeting the objectives. The timing of the goals is typically one year.

Communicating the Strategic Objectives to the Employees and Parent Organization

The Olive Hotel needs to commit to operate responsibly and transparently to its stakeholders and the community. Employees of the hotel are critically important internal stakeholders and should understand the importance of their contributions and their role in achieving the shared vision and shared values of the Olive Hotel. This is accomplished through a planned, transparent, ethical, and socially responsible approach.

- *Employees*: Strategic objectives should be continually communicated to employees through an ongoing internal stakeholder engagement process:
  - Hold regular meetings where employees can share their interests and express concerns
  - Hold regular meetings where management shares its decision making with employees regarding changes in objectives, goals, processes, operations, and policies
  - Hold periodic meetings to report to employees on actions taken on previously expressed concerns, progress toward shared objectives, and any regulatory issues that may arise
  - Hold special meetings devoted to addressing critical events or employee issues
  - Invite employee feedback at any time through face-to-face meetings with management
  - Continually evaluate the stakeholder engagement process for effectiveness
- *Management*: The Accor Group is an international company accustomed to receiving updates to the mission and strategic objectives of its affiliated properties. The following communications

procedure will ensure that Accor and the Olive Hotel are operating as planned:

- Inform Accor prior to undertaking an update of the hotel's mission statement or strategic objectives by providing an outline of the process the hotel will follow
- Send Accor any proposed changes before they are adopted and distributed to stakeholders
- Transmit to Accor final documents as part of the process by which they are disseminated to stakeholders

Internal and External Context (Chapter 8)

Like all organizations, the Olive Hotel operates in an uncertain world. By scanning the internal and external operating environment (i.e., context), the hotel can identify factors that create opportunities and threats for its operations. These are the effects of uncertainty. Successful management of uncertainty leads to a lower risk that the organization will meet its strategic objectives.

Internal Context

The internal context is anything within the Olive Hotel's control or sphere of influence that can affect its ability to meet its objectives. This includes management, organizational structure, hotel policies and operations, employees, relationships with organizations in its value chain (e.g., contractors, logistics, and suppliers), and the hotel's organizational culture. A TECOP analysis method can be used to identify the influences in the organization's internal context. These influences include the following categories: technical, economic, cultural, organizational, and political/regulatory. The TECOP structure was developed to be used in the area of project management. Process mapping can help find the factors that pose opportunities and threats within the internal operating environment. This involves some critical thinking and question asking. One method for asking questions for processes and operations is called SWIFT (i.e., structured what ifs technique) analysis.

The scan of the internal operating environment is usually conducted by a group of individuals internal to the hotel organization. It may have some people familiar with the hospitality sector to make sure that the internal participants are not overlooking influences, factors, opportunities, and threats. This team is often directed by a professional facilitator. The scans should be updated on a regular basis or when a change is noted in the internal operating environment.

External Context

The external context is determined by a scan of the external operating environment. A PESTLE analysis can be used to assess the influences and factors that create opportunities and threats and the uncertainty that is associated with each effect. The influences of the PESTLE structure are political, economic, social, technological, legal, and environmental.

Using One of the Six PESTLE Elements

Looking at the environmental influence as an example, it can be determined that water, energy, and transportation are important factors.

The hotel identifies public transportation in the city of Cambridge as an environmental opportunity. Cambridgeport has nine bus routes, but only one subway stop at its northernmost periphery. Rail is typically preferred over buses by visitors. Parking in Cambridgeport is difficult for visitors and residents alike. There are plans for continued development in the same character and scale as the existing urban fabric. This makes public transportation even more important in the future.[20,21]

As a mixed industrial–residential community of 200+ years, Cambridgeport has aging infrastructure and building stock that is costly to preserve and renovate due to its inherent energy inefficiencies.[22] The Olive Hotel is housed partly in a 120-year-old structure and partly in a 15-year-old structure. Investment in upgrades should be anticipated. However, as an affluent community,[23] Cambridgeport is a relatively safe environment for visitors, and therefore a stable long-term investment for the Olive Hotel. The hotel fills a unique niche in the city of Cambridge. There are three other hotels in Cambridgeport, which are all very large high-rise properties with meeting facilities, two with just more than 200 rooms each and one with 470 rooms.[24]

The city of Cambridge has adopted very ambitious greenhouse gas emissions reductions by 2020 and 2050, which on the one hand makes it a credible sustainability partner for the Olive Hotel and the Accor Group, but on the other hand will drive up energy costs as the city obtains its energy from more clean and renewable energy sources. Finally, clean water is becoming a precious resource worldwide and the Olive Hotel should anticipate increased water costs in the future.

Selecting Factors in the Environmental PESTLE Influence

A simple matrix can be used to record the factors that affect the hotel's ability to identify its opportunities and threats within each influence of the PESTLE analysis. Some factors will overlap several influences. It is more important to identify the factors for the generation of the opportunities and threats than to get them in the correct influence category (i.e., PESTLE). Once the factors are identified, questions and critical thinking are used to find the opportunities and threats associated with each influence–factor combination. Opportunities and threats are both "effects of uncertainty" in the definition of risk: risk is the effect of uncertainty on meeting strategic objectives. By managing the opportunities and threats, the risk will be on the upside or downside of meeting the organization's objectives. Lowering the level of uncertainty is key to successful risk management. Some factors may have both opportunities and threats associated with them.

Based on this assessment of the environmental influences in Cambridgeport, the Olive Hotel should further identify the relevant opportunities and threats as follows:

- Gather data for the amount of current and proposed parking in Cambridgeport
- Investigate opportunities for the Olive Hotel to valet guest vehicles off-site
- Evaluate the Olive Hotel's mechanical equipment against current and anticipated performance standards and energy costs
- Evaluate the space allocated to the hotel's laundry operation against optimal conditions for its employees and operations, as well as against the hotel's other space needs
- Gather data for water consumption by the Olive Hotel compared with similar hotels in the area

Based on the unique qualities of Cambridgeport, two environmental factors that seem relevant to the Olive Hotel's uncertainty analysis (i.e., part of the risk management program) are parking and infrastructure equipment efficiencies. Remember that this case is just providing a means for understanding how to work with the process of scanning the external operating environment. An actual program must look at all the PESTLE influences and develop a method to have a valid list of factors, followed by the means to identify the opportunities and threats.

Three Opportunities as an Example

Based on the PESTLE analysis, environmental factors having to do with parking and resource consumption present three identifiable opportunities:

1. *Valet parking*: The Olive Hotel's lack of parking most likely drives customers to the other three large hotels in Cambridgeport, in spite of its uniqueness as a small hotel with historic character in a vibrant area. The hotel can create an opportunity by offering valet parking at a competitive rate in partnership with a nearby garage. The valet service may not generate revenue, but it will attract customers.

2. *Replace aging mechanical equipment with multifunction heat exchanges*: Equipment at the end of its useful life presents an opportunity to upgrade with minimal loss, and heat domestic water using air condition waste heat, requiring no backup water heating in the summer or colder months.

3. *Outsource the hotel laundry*: Opportunities include converting the space to storage or an exercise facility; avoiding the purchase of expensive water-saving technology, which is a better investment for a larger commercial laundry operation;[25] and avoiding the expense and threats associated with potential upgrades to a safe, pleasant, and fully Occupational Safety and Health Administration (OSHA)–compliant operation.

Three Threats as an Example

Based on the PESTLE analysis above, environmental factors having to do with resource consumption present the following three identifiable threats:

- *High equipment and lighting replacement costs* due to stricter local performance requirements arising from the very ambitious greenhouse gas emissions—more costly than conventional products.
- *Energy costs may rise precipitously* as the city of Cambridge strives to meet its ambitious greenhouse gas emission goals for 2050, described above, as it likely starts procuring more expensive renewable energy to shift its energy profile away from oil and coal, which currently represents 60% of its energy sources.
- *High replacement costs for ultrahigh water-efficient fixtures and equipment*: When the hotel renovates, likely within the next 10 years, it will face the decision and expense of replacing all plumbing fixtures with new expensive water-saving laundering technologies, and possibly upgrading its laundry facilities to attract and retain quality employees.

Later, in the "Uncertainty Analysis of Opportunities and Threats" section, the internal and external opportunities and threats will be rank ordered using uncertainty analysis as a key step in creating a sustainability plan for the Olive Hotel.

Engagement of Stakeholders and Social License to Operate (Chapter 9)

The engagement of stakeholders is a key element of all sustainability programs. These stakeholders are associated with the scans of the internal and external operating environments. Potential stakeholders are identified and thoughtfully analyzed in order to compose a meaningful stakeholder engagement process. If the stakeholders are not involved in an effort to engage with an organization, there could be some "interests" that have not been addressed by the organization that may lead to actions to get the attention of the organization. These actions are prevented with some level of acceptance of the organization as a result of the engagement process. This acceptance is referred to as a "social license to operate."

Selecting External Stakeholders

By scanning the external operating environment, the Olive Hotel can identify external stakeholders. Many organizations create a list of the external stakeholders and a chart that details their interests, influence in the community, relevance to what the local residents believe, and other information needed to initiate the engagement process. The scanning of the external operating environment is repeated on a regular basis or whenever there is a change in the external context. This enables the hotel to manage the effects of uncertainty (i.e., opportunities and threats) so that it can obtain its social license to operate and have a clear path toward meeting its strategic objectives. Using the following process, the Olive Hotel can identify stakeholders according to their interests in the opportunities and threats determined in the determination of the context (i.e., parking and water and energy conservation):

- Stakeholders who may have legal rights—Middlesex County and the city of Cambridge governments, Cambridge and Cambridgeport civic groups, and electric utility providers
- Stakeholders who are most affected—neighbors, Accor Group management, hotel guests, and hotel suppliers
- Stakeholders who are less affected but still have an interest—local advocates; competitor hotels; area hotels, whether competitors or not; lawyers; and the media
- Stakeholders who are less affected and have a moderate or low level of interest—community institutions such as churches and schools, and area hotels of similar character as the Olive Hotel
- Stakeholders with a high level of influence over the hotel's ability to meet its objectives—local governments, local activists, Accor Group management, and guests

Once the key stakeholders are identified, the hotel can create a stakeholder analysis table identifying each stakeholder, their interests and influence, and any additional opportunities and threats that may affect the ability of the hotel to meet its strategic objectives. This information can be organized on a stakeholder analysis template.[26]

Engaging the Stakeholders

Building effective relationships with external stakeholders requires a commitment to transparency and giving equal weight to stakeholders' varying interests so the hotel can earn its social license to operate in an uncertain world.

Although some of the stakeholders will be identified as having higher levels of interest and influence than others, the Olive Hotel will engage all stakeholders equally to give equal weight to their varying interests and earn its social license to operate in a world of uncertainty, especially in terms of resource availability. The hotel must also take into account that some stakeholders will be interested in the hotel's internal risk management and sustainability programs. A suggested engagement process involves the following:

- Individually engage stakeholders with legal jurisdiction over the hotel with meetings, documented correspondence, and public forums, as required.
- Individually engage stakeholders who are most affected, as their interests are typically not common to one another. Engage in individual and group dialogue through meetings and documented correspondence where appropriate, maintaining a high level of transparency.
- Individually engage stakeholders with strong, yet divergent, interests and stakeholders with a high level of influence. Include these stakeholders in groups and public forums pertaining to their interest; however, give special effort to personal engagement to prevent their interests from becoming "issues" that might interfere with the hotel's ability to achieve its objectives.
- Hold community meetings for stakeholders with moderate interests, and apt to be more supportive of the hotel's objectives if they feel that their concerns are being heard. Meetings should be supplemented with written group notifications, as appropriate.
- Following the model of the Accor Group, maintain a high level of transparency and engagement at all levels with employees, guests, corporate partners, and parent organizations; suppliers and service providers; the host community; and all stakeholders who share a deep commitment to environmental stewardship.[27]
- At all levels, the engagement process will be face-to-face whenever feasible and ongoing over time. The process should seek input from stakeholders in designing how they participate, provide the information they need to participate, communicate how their input and contributions affected the decisions of the hotel, and be evaluated regularly for effectiveness and modified for continual improvement.

- Integrate the use of digital media into the engagement process to keep communications timely and fluid.

Managing the Stakeholder Effort

Effective engagement requires a structured process approach that involves planning, allocating adequate resources, implementing, monitoring, measuring, reporting, and taking actions to continually improve the process.[28] The manager of the Olive Hotel is ultimately responsible for the success of stakeholder engagement. Components that could be included in the manager's efforts include

- *Stakeholder coordinator (SC)*: Assign a single individual who has responsibility for the coordination of communication and information. This individual will be the day-to-day "face" of the Olive Hotel with both internal and external stakeholders, and should have a high level of intelligence and interpersonal skills.

- *Process maps*: Management, with the SC, shall develop two sets of process maps, one for the internal and one for the external stakeholders, that identify the engagement procedures and responsible individuals. The SC should also create standard forms for documenting all conversations, process notes, and outcomes.

- *Contact database*: The SC should create a database of all known and potential stakeholders. The database should have maximum sorting capabilities.

- *Communication database*: The SC should create a sortable database in which all conversations are logged and linked to file notes for each conversation and action.

- *Stakeholder calendar*: The SC should create a sharable calendar for coordinating all face-to-face and telephone meetings. The calendar should be regularly backed up and added to file notes.

- *Filing system*: The SC should create hierarchical paper, digital, and email filling systems that are organized, clear, and easy to navigate, search, and archive. Master indices should be created for each of the three filing systems. A consistent file naming system should be used.

- *Communication*: Regular, detailed verbal and email communication should occur between the SC and the hotel manager. In addition the SC should prepare periodic reports, the frequency of which shall be decided in consultation with the hotel manager, that summarize the ongoing stakeholder engagement efforts and results. These reports will be used to summarize results, evaluate the process for efficiency and effectiveness, and inform the Accor Group of the Olive Hotel's stakeholder engagement efforts. The manager would use all communication with the SC and stakeholders to determine the methods, timing, and participating individuals for each form of engagement, for the most efficacious process.

Communicating the Engagement Information with Management

With 32 employees, the Olive Hotel is small enough for regular ad hoc communication to occur between the hotel manager and the SC. All communications, no matter how informal, should be documented and filed according to the procedures outline above. In addition, results of the hotel's stakeholder engagement process should be communicated regularly and transparently to its parent organization, the Accor Hotel Group. Communication with the Accor Group would occur primarily through the standard reports generated by the SC and signed by the hotel manager. In light of Accor's commitment to regular, transparent dialogue with shareholders and partners,[29] both the hotel manager and SC may find that more frequent and informal communications occur. It is imperative that all communications, no matter how informal, are documented and filed according to the procedures outlined above.

Social License to Operate

The social license to operate is defined as existing when an organization has ongoing "approval" within the local community and from other stakeholders. The purpose of stakeholder engagement is to obtain and maintain the social license to operate in the community and larger external context by building legitimacy, credibility, and trust between the hotel and its stakeholders. Legitimacy is built on genuine understanding of stakeholder interests and the transparent exchange of information. Credibility is based on consistently providing accurate information and honoring all commitments made to the stakeholders. Trust comes from an effective engagement process characterized by acknowledgment and respect for all stakeholder interests, shared experiences, and collaboration. It is important to remember that although the hotel must follow certain practices to earn the social license to operate, the stakeholders and community grant the license not on the basis of engagement activities, but on the basis of the quality of the relationship.

Three Things That Help Secure the License to Operate

There are three things that can be done by management to assist the Olive Hotel in securing its social license to operate:

1. *Be a social purpose leader*: The senior management identifies and embraces a social purpose within the community so the Olive Hotel becomes an accepted important part of the community.

2. *Give more control to local communities and stakeholders*: The senior manager should create collaboration areas where interaction is easy and welcome. Stakeholders feel that they have been heard, so that the likelihood of a conflict and disagreement is reduced and the hotel's credibility is enhanced. Good community relations can make it easier to attract workers, as employees want to work for a hotel that is respected within the local community. Additionally, good connections with the community create a transparent stakeholder process because of deeper insight into the external context of the hotel.

3. *Build partnerships with community governments and organizations*: Working with community entities requires an investment of time, particularly at the beginning of the relationship, as each side gets to know and understand the interests, values, and options of the other. As with any core business activity, the goal is that this initial investment of resources will benefit the Olive Hotel in future times.

Maintaining the Social License to Operate

Three things that can be done by management to maintain the social license to operate over time are as follows:

1. It is important that the senior manager does not take the license to operate for granted. The social license to operate is dynamic, as peoples' perceptions, beliefs, and opinions change over time for many different reasons.

2. Establish network and strategic partnerships with other organizations in the Cambridge area and participate in key community social events.

3. Maintain a consistent, regular flow of communication with stakeholders so concerns and issues are readily identified. An open and ongoing communication can improve the relationship quality between the hotel and the local community and develop mutual respect, inclusion, honesty, plain disclosure of information, and transparency in the hotel's strategies, processes, activities, and operations.

Additional information regarding the social license to operate is addressed in the "Measurement, Transparency, and Accountability" section in Appendix III.

Organizational Governance and Leadership (Chapter 10)

Governance is the system by which the hotel makes and implements decisions in pursuit of its strategic objectives. Therefore, governance not only is a component of the hotel's sustainability plan, but also has the mandate and responsibility to implement itself as a component of the sustainability plan.

It is important that the mandate of governance extend to every component of the sustainability plan. To illustrate, one should look at the relationship between governance and engagement with stakeholders. Stakeholder engagement is the process by which several important objectives are met. First, it is the means by which the hotel understands its external operating environment, which forms the basis for identifying and managing the opportunities and threats that affect its ability to meet its objectives. Second, it provides active sense making and knowledge, which help the hotel establish a competitive advantage over the other hotels in Cambridge and the surrounding area. Third, it drives the need for transparency and accountability that builds legitimacy, credibility, and trust with employees and the community. Fourth, it is the process by which shared values are created with the community that help the hotel attain and maintain its social license to operate.[30] These are all management functions. It is the governance and leadership that provide the structure and integration by which all components of the sustainability plan are implemented. Following are some examples of good governance and leadership practices that the Olive Hotel would implement as part of a sustainability plan.

Leadership Processes

The following sustainable leadership practices, when implemented by the hotel manager, not only enhance the performance and effectiveness of the hotel's stakeholder engagement process, but also lay the foundation for the hotel's standards of behavior as stated in its code of conduct:

- Enthusiastically participate in the development of the hotel's stakeholder engagement plan
- Build trust with both internal and external stakeholders by implementing a culture of ethical behavior, ownership, involvement, continual improvement, transparency, and management accountability—all values that are in alignment with Accor Group's governance
- Set an example for integrity, respect for stakeholder interest, human rights, and legal obligations
- Promptly address conflicts of interest between the stakeholders and the hotel

Establishing the Sustainability Policy

After the hotel develops its sustainability plan, it will need to create a sustainability policy for implementing the plan. The policy sets the guidelines for the basic steps under which procedures of the plan will be developed. It is the responsibility of leadership to create and maintain the policy. It is used initially to introduce the sustainability plan to the internal and external stakeholders and invite questions and concerns. There might be revisions to the policy to address significant concerns before the final plan is initiated. The policy should be sent periodically to stakeholders and be prominently displayed in the hotel where employees can access it. When there is a change in the sustainability plan, the policy should be revised and redistributed to all stakeholders, accompanied by an explanation of the changes and how the changes will manifest themselves in the operations of the hotel. The following information should be included in the sustainability policy:

- An easily identifiable title with the date it was implemented
- Description of what the policy is and how it is meant to be used
- The purpose for which the sustainability plan exists
- The foundational principles that form the basis for the sustainability plan
- The system of processes and procedures that will be followed in implementing the sustainability plan
- The type of process that will be utilized
- Who is responsible for executing the sustainability plan

An example of a sustainability policy for the Olive Hotel can be found in Box II.2.

Uncertainty Analysis of Opportunities and Threats (Chapter 11)

Like other components of a sustainability plan, uncertainty analysis follows a process approach in order to embed risk-based thinking into all operations and all decisions every day. The uncertainty assessment is the core process the hotel uses to manage the effects of uncertainty. Uncertainty is characterized by factors, opportunities, and threats that come from the scanning of the internal and external operating environment. The framework for determining the significant opportunities and threats is similar to that used in risk assessment. It consists of

- Defining the uncertainty criteria
- Establishing the uncertainty context
- Stakeholder engagement

BOX II.2

It is the policy of the Olive Hotel to meet our strategic objectives in an uncertain world using the 11 principles of sustainability outlined below. Sustainability is defined as the ability of the Olive Hotel to transparently manage our responsibilities for environmental steward-ship, social well-being, and economic prosperity through shared value with our stakeholders over the long term while being held accountable by our stakeholders.

We will use the following principles to guide our way to sustainability:

- Sustainability creates and protects value for all our stakeholders.
- It is integral to how we operate every day.
- It is embedded within the way we make decisions.
- It helps us address uncertainty.
- It is a systematic, structured, and timely consideration that guides all activities.
- It is based on the best information from the knowledge of our members and the collective wisdom of our external stakehold-ers and value chain.
- It is tailored specifically to the Olive Hotel to help us meet our objectives.
- It takes human and cultural factors into account.
- It is dynamic, interactive, and responsive to change in our external operating environment.
- It is transparent and inclusive with our engagement with stakeholders.
- It is used by the Olive Hotel to innovate so we can continually improve in our quest to meet our strategic objectives.

- Uncertainty assessment
- Uncertainty analysis
- Uncertainty response and plan
- Monitoring and review

Uncertainty criteria are the terms of reference against which the signifi-cance of an opportunity or threat is evaluated. The criteria are based on the Olive Hotel's strategic objectives, and the likelihood and consequences of the opportunity or threat occurring (see the "Uncertainty Analysis: Assigning Likelihoods and Consequences" section). The criteria must include terms for expressing the level of uncertainty and rules for deciding the significance of the opportunity or threat. Uncertainty criteria for evaluating an opportunity or threat facing the hotel might include the cost of addressing the opportunity or threat, its effect on employee morale measured in terms of absenteeism, and its potential impact on bookings as a percentage increase or decrease in revenue.

Uncertainty context refers to the "source of uncertainty" related to the objectives or decisions particular to the opportunity or threat being assessed. Consequences, relevant operating factors, related opportunities and threats, and stakeholder involvement are all part of the uncertainty context. For example, if the opportunity being assessed is reducing water consumption, the uncertainty context might include deciding which areas of water consumption to measure and what measurement methods to use, the cost of special personnel or equipment that may be required, and whether there is stakeholder interest in conserving water in particular areas.

Stakeholder engagement refers to the stakeholders who are involved in the opportunities or threats being assessed. They may be involved through their interests in the hotel's policies, operations, or environmental impact. The stakeholders in the hotel's water reduction would include Accor, employees, guests, and concerned citizens.

Uncertainty assessment is the core process for determining how to manage uncertainty. It is broken down into the subprocesses of uncertainty identification, analysis, and evaluation, which are exemplified in the following subsections using the three opportunities and three threats identified in Chapter 8 (context).

Uncertainty response and management are designed to modify processes or activities in order to influence the outcome of opportunities and threats in the internal and external operating environments. Determining whether to respond to an opportunity or threat that was evaluated as unacceptable involves balancing not only resources required against benefits derived, but also consequences against likelihood. For example, a threat with severe consequences might have an extremely small likelihood of occurrence, causing response to appear unjustified. However, if the consequences threaten the hotel's social license to operate, the response might be warranted. High-impact outlier occurrences like this are sometimes referred to as "black swans."[31] Selecting responses requires a process approach that includes defining options, proposing actions, allocating resources, engaging stakeholders, an uncertainty assessment of the response, and assessing the effectiveness and efficiency of the response so it can be improved, if necessary.

Monitoring and review of all processes involved in the six components of uncertainty management is necessary to determine the effectiveness and efficiency of the system, identify areas for improvement, increase the hotel's knowledge of its internal and external context, and determine whether the hotel's resources were put to the best use.

Uncertainty Identification for the Olive Hotel

Uncertainty identification entails compiling a list of all the factors, opportunities, and threats that might help or harm the achievement of the hotel's strategic objectives. The list should include the factors, opportunities, and threats identified in the internal and external context using the PESTLE and TECOP influences in Chapter 8 (context). It should be as comprehensive as possible so as not to miss a serious threat or significant opportunity. An identification tool, such as an uncertainty register, may be used to assist in identifying sources; causes, such as an event, decision, or change in context; underlying causes; existing operational controls; the timing, likelihood, and consequences of an occurrence; who might be affected; and the possible outcomes. It should also include cascading or cumulative opportunities and threats resulting from other factors, opportunities, and threats. Documentation of uncertainty identification should include the scope of what was identified, the methods, the uses, the participants, and the date.

Uncertainty Analysis: Assigning Likelihoods and Consequences

Uncertainty analysis involves analyzing the nature of opportunities and threats in terms of their likelihoods and consequences. Consequences are the outcomes of opportunities and threats, and likelihoods are the chances of them happening. Brainstorming the likelihoods and consequences for the identified opportunities and threats, and assigning values to them, would be led by the hotel management, and should also include groups of various internal stakeholders who can bring a diverse perspective to the analysis.

The key opportunities involve valet parking; heating, ventilating, and air-conditioning (HVAC) equipment upgrades; and the outsourcing of the laundry. Key threats involve HVAC and lighting upgrades, increased energy supply costs, and water efficiency upgrades. The qualitative uncertainty analysis outcome can be quantitatively determined based on the compilation of the likelihood and the consequence of each opportunity and threat. The analysis of the likelihoods and consequences for the three opportunities and threats identified in the context section can be conducted using the uncertainty matrix.

Uncertainty Evaluation

An uncertainty evaluation uses the results of the uncertainty analysis to assist in making decisions about appropriate responses to the key opportunities and threats. Based on the combined numeric values assigned to the likelihoods and consequences, a level of risk or risk priority can be assigned to each opportunity and threat. Opportunities and threats can be evaluated simultaneously to identify areas where opportunities might offset threats by using a bipolar scale where the consequences and threats are represented by negative numbers and consequences of opportunities are represented by positive numbers. A "mirror-image risk map" is used to place the most positive opportunities in close proximity to the most negative threats. It is helpful to have a visual tool for evaluating possible situations where the development of an opportunity can offset a threat.

A risk priority number is used to identify the possible combinations or targets to respond to.

- The major negative threat of high-cost upgrades for energy-efficient mechanical equipment and lighting is offset by the major positive opportunity of energy use reductions.
- The major negative threat of the higher cost of energy supplied by the utility companies is offset by the major positive opportunity of energy use reduction.
- The major negative threat of high-cost upgrades for water-saving equipment and fixtures is offset by the extreme positive opportunity of water savings from the new equipment and fixtures, and potentially from the outsourcings of the laundry whereby new advanced laundry equipment will not have to be purchased.
- The major positive of valet parking offsets all threats by potentially creating revenue from increased guest bookings and preserving the unique niche that the Olive Hotel fills in Cambridge as a small hotel with historic character in a vibrant area.

It is important to note that while the major positives and negatives appear to offset each other, even if all solutions are implemented, the hotel will still incur high costs. It is not anticipated that resource savings will offset the anticipated expenses of physical plant upgrades and higher energy costs. The threats are therefore assigned an extremely high uncertainty priority, while the opportunities are assigned a very high uncertainty priority. The results of this uncertainty evaluation, including the offsets described above, will determine the response by the hotel leadership that will best manage the opportunities and threats and help the hotel meet its strategic objectives.

Documenting the Uncertainty Analysis of an Opportunity

The opportunities identified above (valet parking, HVAC equipment upgrades, and laundry outsourcing) can create benefits for the hotel; however, they are risky. They require proactive efforts to create, and there are consequences to both failure and success. Documenting the uncertainty analysis of these opportunities should include

- Key assumptions and limitations used to define an opportunity
- Sources of information used to identify the source of each opportunity
- Existence of programs to enhance the likelihood of an opportunity succeeding
- Explanation of the analysis method and the definitions of the terms used to specify the likelihood and consequences of each opportunity
- Description of the opportunities and the potential benefits
- Likelihood and consequences (positive and negative) of these specific occurrences
- Resulting level of uncertainty
- Effect on the level of uncertainty (opportunities with a high level of uncertainty should be flagged for future review)
- Response plans for execution of the opportunities, including exploiting, enhancing, sharing, or ignoring each opportunity

Documenting the Uncertainty Analysis of a Threat

While the procedure for documenting the uncertainty analysis of the threats (HVAC and lighting upgrades, increased energy supply costs, and water efficiency upgrades) is similar to that of opportunities, the consideration of how to act is fundamentally opposite, as described in the list below. Documentation should include

- Key assumptions and limitations used to define a threat
- Spruces of information used to identify the source of each threat
- Existence of controls for threats and their effectiveness
- Explanation of the analysis method and the definitions of the terms used to specify the likelihood and consequences of each threat
- Description of threats and severity of consequences
- Likelihood and consequences of these specific occurrences
- Resulting level of uncertainty
- Identified tolerability of each uncertainty
- Effect on the level of uncertainty (threats with a high level of uncertainty should be flagged for future review)
- Records of low-priority uncertainties that were screened and why they did not progress
- Response plans for threats, including avoiding, transferring, mitigating, and accepting each threat

Organizational Operating Systems (Chapter 12)

The Olive Hotel's operating system is made up of processes and specific operations. Operations consist of a set of processes and the leadership, facilities, and resources necessary to conduct the processes by which the hotel meets its strategic objectives. The processes are further broken down into a collection of activities performed by people, aimed at delivering products or services. In the case of the hotel, the service it delivers is overnight accommodations. All operational activities require resources—financial, material, human, knowledge, and natural. These resources must be used effectively and efficiently if the hotel is going to meet its environmental, social, and economic sustainability responsibilities. It is therefore essential to evaluate all processes from the perspective of how they consume resources. Since the hotel laundry is an intensive user of resources, we will look at how the hotel can make its laundry processes effective, efficient, and efficacious, with the goal of consuming the least possible amount of resources.

Process Diagram for the Laundry Operations

In order to examine the laundry operations, it is helpful to first create a process diagram for the activities that make up those operations. The process map is a visual representation of the flow of activities in the hotel laundry. The diagram will serve as the foundation for continuous quality improvement because it is based on the operational objectives selected in Chapter 7: a suitable and legal work environment, resource efficiencies, and guest participation. Using a pictorial format in the process map allows the hotel to examine its process so that it can discover threats to address and opportunities for improvement, establish clear accountability, facilitate communication, prevent regulatory violations, identify environmental uncertainty and occupational health and safety hazards, and conform to the process approach requirements of a sound system of management.

A diagram illustrates the full operation cycle, beginning with the guest-ready room. The staff stocks the room with clean linens and ensures that guest conservation measures are in place and invitingly presented. When a room is vacated, soiled linens are safely collected and transported to the laundry facility, via a laundry chute if possible. In the laundry facility, linens are safely sorted by type and examined for wear and stains. Linens are then washed, dried, pressed, and folded. The hotel must select pressing and folding equipment based on need and available resources. Clean and folded linens are then sorted, stocked on carts, and transported to the guest rooms in preparation for the next occupants.

Maintaining an Effective Laundry Operation

Effective processes are those that are the correct ones for satisfying operational objectives. Maintaining an effective laundry process means that the process is the correct one for delivering what is required by the hotel's guests and stakeholders. Even if the Olive Hotel's laundry process is efficient, as defined by the process, it will not be effective if it is not the correct process, or not "as correct" as the hotel could employ. The laundry process, and all processes in the hotel, must be effective if the hotel is going to meet its objectives. The following procedure will help the hotel achieve and maintain an effective laundry process:

- Define what the hotel requires of its laundry operation, both qualitatively and quantitatively.[32] Consider the hotel's mission and strategic objectives to identify qualitative requirements, and internal and external resource needs to identify quantitative requirements.
- Use the comprehensive uncertainty analysis and sustainability management tools described in Chapter 11 to prioritize opportunities and threats so that resources are most effectively allocated to satisfy customers and stakeholders.
- Use a plan–do–check–act (PDCA) process as a means by which the laundry operation can be monitored and improved for greater effectiveness.

Maintaining an Efficient Laundry Operation

Efficient operations are the by-product of effective processes. As processes become more effective, the operations use fewer resources to achieve the intended outcome and they become more efficient. A system of efficient operations is critical for the hotel's sustainability program, the purpose of which is to help it meet its objectives in an uncertain world. By applying a process approach[33] to conserving resources, resource management becomes the process by which operations become more efficient, that is, "do more with less." The following resource management approach will help the Olive Hotel focus on the effective use of resources in order to improve the efficiency of its laundry operation.

- Define the activities needed to create an effective laundry operating process.
- Determine how the laundry processes interact with other processes, such as food service, housekeeping, procurement, and facility management.
- Define the resources required for the laundry operation. Resources used by the laundry include financial (cost of utilities, salaries,

products, and facilities), human (employees and suppliers), competency (skills provided by employees and suppliers), infrastructure (physical plant required to operate the laundry), work environment (quality of the space, air, light, safety, security, and suppliers), technology (equipment, detergent, and stain removal products), and natural (water, electricity, natural gas, and raw materials that go into laundry products and the resources embedded in obtaining them) resources.

- Identify interdependencies between processes to see where resources can be shared. For example, the hotel might be able to use heat recovery from HVAC equipment to preheat water for the laundry operations.
- Define responsibilities and authorities for the laundry process.
  - Assign responsibility for implementing the laundry process; for example, designate managers and supervisors.
  - Consider giving the laundry process authority over resource matters in other operational processes; for example, laundry management can implement procedures for food service and housekeeping to pretreat stains, thus reducing the cost of "linen loss" due to staining.
- Define how the laundry processes will be documented by determining what will be documented and how it will be documented; for example, the laundry operation might display the process map as a framework for identifying opportunities and threats and determine what to monitor and measure so it can meet its operational objectives.
- Assess the uncertainty associated with making the laundry operation more efficient. Reducing resources in laundry processes can create uncertainty by making the processes less effective. For example, moving to laundering technologies that use less water might make the linens appear less clean, affecting guest satisfaction. Before reducing water consumption in the washing process, an uncertainty assessment should be conducted. This will help assess whether the washing processes can be made more resource efficient without creating a threat to the operations and to the ability of the hotel to meet its objectives. Resource scarcity and fluctuating resource costs can also pose threats and should also be considered in the uncertainty assessment. Once processes are identified for resource reduction, the proper approach must be determined to ensure the continued or improved effectiveness of the processes and the entire laundry operation. Perhaps there are processes in place that can reduce resource use. These should be identified. Water resource reduction (or scarcity) can interrupt an otherwise efficient laundry operation, so the implementation strategy must take this into consideration by

planning alternate means of ensuring that the laundry gets done effectively during the implementation.[34]

- Define the monitoring and measurement requirements for the laundry processes.
  - Determine where and how monitoring and measuring should be applied. For example, if the laundry wanted to monitor its water usage, how should it do this? Should water usage be submetered? Tracked in some other way? Who will monitor the usage?
  - Determine how to monitor for both control (lead measures that drive productivity) and improvement (lag measures that document past performance).
  - Determine monitoring and measurement requirements. For example, what equipment, people time, and revenue resources are needed to monitor water usage, and what is their availability? Stakeholders may ask for resource consumption and savings data, particularly for natural resources that are shared with society.
  - Determine how results can be recorded in a consistent manner for all processes.
- Quantify the resources needed to ensure that the laundry process is effective.
- Verify the effectiveness of the laundry operation against its planned objectives and implement changes for improvement.
  - Are the laundry process objectives consistent with the hotel's strategic objectives?
  - Has the hotel effectively and efficiently embedded resource management into other objectives in its operations?
  - How has the resource management approach to creating efficient laundry operations contributed to meeting the hotel's sustainability outcomes and strategic objectives?
  - How does an efficient and effective hotel operation create value for the Olive Hotel?

Maintaining an Efficacious Strategy for the Laundry Process

The term *efficacious* is widely defined as having the power to produce a desired effect or outcome, which is the case of the laundry it to help the hotel meet its strategic objectives in an uncertain world. The prerequisites for an efficacious strategy are efficient and effective processes. An efficacious strategy is created by the following procedures:

- Define the purpose of the laundry and identify three or four clear operational objectives, for example, embed hotel sustainability

outcomes into operations, maintain high standards of linen cleanliness, achieve the optimal balance between laundry productivity
and linen inventory, and create a healthy work environment for
employees.

- Identify any factors in the hotel's external context that might impact
  the laundry's operational objectives, for example, municipal water
  rationing in the event of a severe regional drought.
- Identify opportunities and threats in both the internal and external
  context that might impact the laundry's operational objectives, for
  example, a water main break.
- Engage internal stakeholders to help identify uncertainty and create support and goodwill for the operational objectives. Employees
  can provide valuable feedback, such as the need to raise the height
  of laundry carts to improve worker comfort, and therefore increase
  productivity and goodwill.
- Conduct an uncertainty assessment to ensure that operational decisions will reliably deliver the desired strategic outcomes of effectively managing uncertainty and optimize the consequences of
  uncertainty; for example, uncertainty assessment can be used to
  determine whether the interests of hotel guests or the interests of
  neighbors have more influence over the hotel's ability to meet its
  objectives.
- Establish individual goals for achieving the operational objectives
  of the laundry; for example, if an operational objective is to reduce
  water consumption, identify how specific individuals in the laundry
  operation will effect water reduction through setting relevant goals.
- Create action plans for achieving the stated goals in a given time
  frame.
- Outline a process for monitoring and evaluating the effectiveness
  of the above strategy and for modifying the strategy for continual
  improvement.

Organizational Support Operations (Chapter 13)

The Olive Hotel is in the business of providing services to people by people.
The hotel's mission is *to provide affordable and healthy accommodations in a comfortable and friendly setting*. Every person who works at the hotel plays a role
in delivering affordable, healthy, comfortable, and friendly services to the
hotel's customers. Employees are a critical resource to the Olive Hotel. A key
to maximizing this resource is to provide a work environment where people

feel a shared sense of purpose and culture that encourages a high level of involvement in fulfilling the hotel's mission and strategic objectives.

The leadership of the Olive Hotel should make it a priority to invest in its human capital (i.e., the physical, intellectual, emotional, and spiritual capacities of its people) by cultivating an environment of open communication, trust, stability, and employee development. Such an environment builds mutual understanding and employee satisfaction, and consequently leads to increased employee motivation, creativity, and collaboration. Within this internal context, the hotel's leadership should set strategic objectives that encourage learning, ethical behavior, respect for stakeholder interests, and risk awareness. These strategic objectives, if effectively communicated and affected through the hotel's processes and operations, will become guiding principles that will help to form an organizational culture where everyone engages in innovative responses to environmental, social, economic, and competitive opportunists and threats. These collective endeavors will result in a more resilient and sustainable Olive Hotel.

Creating a positive organizational culture that gives rise to sustainability requires a process approach, as presented in Chapter 12. The operational objective is to transform the hotel's knowledge about its human capital into policies and processes that support its strategic objectives. To develop these culturally consistent policies and processes, the hotel can use a value chain model (Chapter 15) to identify its people-focused "support" processes and the approach, deployment, assessment, and refinement (ADAR) approach (Chapter 16) illustrated in this section to decide how to develop and improve the cultural efficacy of each of these support processes.

Value Chain Model

Operational processes are made up of activities, some of which require actions performed by people. The people-focused activities in the Olive Hotel's operations or "support activities" are found in the organization's value chain diagram. These support activities are in the horizontal rows at the top of the model. Each heading is a description of the people and the activities for which they are responsible. The employee engagement (human resources) and stakeholder engagement activities are involved with the internal and external context. These activities influence and are influenced by the organizational culture of the hotel. The directors of both these activities report directly to the hotel manager, as an indication of their importance to the effective management of the hotel.

Improving the Support Operations

Using the ADAR approach, it is possible to see how the Olive Hotel can improve one of its support activities: the Guest Partnership Program. This program is commonly referred to in the hotel industry as "linen reuse." It involves encouraging hotel guests to use their bed and bath linens multiple times in order to reduce the amount of linens the hotel needs to launder each day. The Olive Hotel's program is in its infancy, and the hotel would like to improve upon the program to conserve water. The issue of saving water was raised by an external stakeholder in regard to the hotel using a "significant" amount of water for landscaping. This led the hotel to propose to reduce water usage, starting with the laundry because of the high amounts of water used in that location. The process by which the hotel arrived at this alternative is discussed in the "Measurement, Transparency, and Accountability" section in Appendix III. Currently, the Guest Partnership Program consists of providing the housekeeping staff with printed cards, instructing guests how to signal that they do not need their towels or sheets to be replaced each day. Spot checks of the guest rooms by management indicate that the cards are often damaged or missing from the rooms. The hotel currently has no method for counting the items that are sent to the laundry, or comparing the laundry volume with guest occupancy.

The ADAR approach breaks down an operational activity into three components: *approach, deployment,* and *assessment and refinement*. Each of these components is briefly defined and then developed below. More detailed descriptions can be found in Chapter 16.

Approach

The approach is the set of processes and activities by which the Guest Partnership Program is executed in a way that helps the hotel become more environmentally, socially, and economically responsible. The intent of the program is to encourage guests to use their sheets and towels for more than one day in order to reduce linen waste and consumption of water, detergent, electricity, and natural gas. This would address the hotel's objective of conserving natural and financial resources.

The current practice of placing cards in the guest rooms explaining to guests that they can contribute to resource conservation by using their linens for more than one day is a good start. The practice should be modified to ensure that the cards are attractive, durable, simple, and clear; visibly displayed in a sturdy and consistent manner; and monitored by management. Management needs to consider that the program might pose a threat to the perceived stability of the housekeepers' and laundry operators' jobs (with less activity, they may need to reduce the working hours of the staff or lay them off). Providing a consistent procedure for displaying and monitoring the signs will shift control of the program away from staff and back to management. Concurrent with stronger controls, the hotel must also invest in open communication, trust, and employee development, so the employees are more likely to feel secure in their jobs and engage constructively in the program and in the strategic objectives of the hotel.

Careful consideration should be given to understanding the values and norms of the hotel guests. There are several ways for the hotel to conserve water, among them being low-flow faucets,[35] waterless or high-efficiency laundry equipment,[36] or outsourcing the whole laundry operation to an outside vendor. In order to give this program the best chance of succeeding, the message has to reflect the values of the hotel guests.[37] A clear message is not enough; it has to produce the desired results. Shaping the message and the modes of delivery so it will have the highest probability of motivating guests reduces the chance that the program will fail because of ineffective messaging. Engage employees from throughout the hotel in brainstorming ideas about motivating guests. The grounds-keeper or the sous-chef might have fresher ideas than the housekeeping supervisor.

The program requires participation by marketing, housekeeping, laundry, procurement, finance, and other operations personnel. The key process steps in implementing the program are listed below by participant.

- *Marketing*: Engage in brainstorming the message—design, vet, and commission the means and media for communicating the program to guests and employees.
- *Housekeeping*: Engage in brainstorming the message—engage staff in brainstorming linen handling procedures to make them effective. Develop action plans and work with staff to set their own goals for successful achievement of the action plans. Train staff. Deploy and monitor activities in the action plans, and track and reward progress.
- *Laundry*: Engage in brainstorming the message—monitor and measure the quantity of laundry each day.
- *Procurement*: Engage in brainstorming the message—monitor and measure the quantity of laundry products and marketing materials used.
- *Finance*: Engage in brainstorming the message—monitor the revenue, number of guest-nights, and cost of utilities and materials.
- *Other operations*: Research linen reuse programs in other hotels for benchmarking. Set a time frame for monitoring the baseline program and the new improved program. Analyze data: utility costs, quantity of laundry, number of guest-nights, costs, and revenue.

Operational objectives for the program should be set by management based on benchmarking from other hotel linen reuse programs and their budget for allocating financial resources to the program. Goal setting should be done by the housekeeping staff based on the action plans they develop for implementing the program.

Deployment

The deployment of the approach described above will provide the details for how the program's activities will be implemented, by whom, and by when. The activities will be organized into action plans that focus on lead indicators that drive performance toward achieving the strategic objective of conserving water and saving money. Employees will participate in setting their own goals for their respective action plans. Effective employee goal setting involves[38] (1) setting two or three "wildly important goals"; (2) identifying "lead" measures that predict and influence success; (3) creating a simple, visible "scoreboard" where employees can track their progress; and (4) meeting often to keep employees thinking about immediate actions they can take to impact the scoreboard.

Management's operational objective, based on benchmarking,[39] is to reduce water consumption in the laundry operation by 17% and break even on the initial marketing and planning costs within one year. These operational objectives support the strategic objectives stated in the program, thereby helping the Olive Hotel remain sustainable over the long term.

The measurable results will be

- The reduction in the number of laundered towels per guest-night over six months, compared with a six-month baseline.
- The reduction in water, electricity, detergent, and natural gas used by the hotel over six months compared with a six-month baseline Note: The hotel does not have submetering in the laundry facility for any of these utilities. The impact of the program on the utility usage by the laundry will be approximated based on utility usage for the entire hotel. It will be important to hold other utility consumption as constant as possible during the one year baseline and test period.
- The reduction in the laundry products used over six months, compared with a six-month baseline. This measure will show whether the amount of laundry product used per item remains constant when items come to the laundry more soiled as a result of being used longer between washings.
- The cost of marketing materials and the planning effort.

Action Plan

For the purposes of this study, the hotel will prepare an action plan to illustrate only the activities associated with signage and linen handling within the guest rooms, and does not include laundry, data tracking, or messaging activities outside the guest rooms. Each activity would have its own action plan.

Information from Chapters 13 and 14 need to be addressed to help implement the action plan.

References

1. Accor Group. (2014). Fact sheet. Retrieved from http://www.accorhotels-group.com/en/group/accorhotels-company-profile.html.
2. Accor Group. (2009). Accor management ethics. Retrieved from http://www.accorhotels-group.com/en/sustainable-development/a-committed-group.html.
3. Pojasek, R.B. (2014). Instructions prepared for the semester paper in the course "Fundamentals of organizational sustainability." Retrieved from http://www.extension.harvard.edu/academics/courses/fundamentals-organizational-sustainability/21808.
4. City of Cambridge. (2013). Cambridgeport neighborhood profile—Neighborhood: Cambridgeport/area 5. Retrieved from http://www.cambridgema.gov/CDD/kplanud/neighplan/neighs/~/media/08A37CA2ADA14F37BF2978E0C028BFFC.ashx.
5. Ibid.
6. U.S. Department of Transportation. (2011). 2009 national household travel survey. Retrieved from http://nhts.ornl.gov/2009/pub/stt.pdf.
7. City of Cambridge. (2008). Climate protection plan. Retrieved from http://www.cambridgema.gov/~/media/Files/CDD/climate/climateplans/climate_plan.ashx.
8. City of Cambridge. (2011). Climate action status report 2011. Retrieved from http://www.accorhotels-group.com/en/group/accorhotels-strategic-vision.html.
9. Accor Group. (2015). Strategic vision. Retrieved from http://www.accorhotels-group.com/en/group/accorhotels-strategic-vision.html.
10. Hotels.com. (2015). Hotels in Cambridgeport in the city of Cambridge, Massachusetts. Retrieved from http://www.hotels.com/search.do?resolved-location=GEO_LOCATION%3ACambridgeport,%20Cambridge,%20MA,%20USA%7C42.35957717895508%7C-71.10771942138672%3AGEOCODE%3AUNKNOWN&q-destination=Cambridgeport,%20Cambridge,%20MA,%20USA&q-rooms=1&q-room-0-adults=2&q-room-0-children=0&sort-order=DISTANCE_FROM_LANDMARK.
11. Accor Group. (2014). Accor ethics and corporate responsibility charter. Retrieved from http://www.accorhotels-group.com/fileadmin/user_upload/Contenus_Accor/Commun/pdf/EN/accor_ethics_csr_charter_2015.pdf.
12. City of Cambridge. (2011). Mission statement. Retrieved from http://www.cambridgema.gov/CDD/econdev/resources/missionstatement.aspx.
13. Accor Group. (2015). PLANET 21 Program. Retrieved from http://www.ccohs.ca/oshanswers/occup_workplace/laundry.html.
14. Canadian Center for Occupational Health and Safety. (2007). Hotel laundry OHS fact sheet. Retrieved from http://www.ccohs.ca/oshanswers/occup_workplace/laundry.html.
15. U.S. Occupational Safety and Health Agency. (2009). Compliance with the OSHA bloodborne pathogens standard. 29 CFR 1910.1030. Retrieved from https://www.osha.gov/pls/oshaweb/owadisp.show_document?p_table=INTERPRETATIONS&p_id=27008.

16. U.S. Environmental Protection Agency. (2015). Hotel challenge. Retrieved from http://www.epa.gov/watersense/commercial/challenge_tools.html.
17. Xeroscleaning. (2014). Near waterless laundry operations. Retrieved from http://www.xeroscleaning.com/blog/emerging-trend-hotels-adopt-near-waterless-laundry-operations.
18. Apogee Interactive, Inc. (2013). Multifunction units for heat recovery. Retrieved from http://c03.apogee.net/contentplayer/?coursetype=ces&utilityid=coastepa&id=1125.
19. Ecolab. (2015). Tackling the toughest challenge in hotel laundries. Retrieved from http://www.ecolab.com/story/tackling-the-toughest-challenge-in-hotel-laundry/.
20. Viggo von Moltke, W. (1986). Cambridgeport Blue Ribbon Committee report. Cambridge, MA: Harvard Graduate School of Design. Retrieved from http://www.cambridgema.gov/~/media/Files/CDD/Planning/Studies/Cambridgeport/cport_blue_ribbon_report_1986.pdf.
21. City of Cambridge. (2003). South Cambridgeport development guidelines. Retrieved from http://www.cambridgema.gov/CDD/zoninganddevelopment/Zoning/designguidelines.
22. Kennedy, M. (2012). Cambridgeport: Its people and their stories. Newtownne Chronicle XI, no. III. Retrieved from http://cambridgehistory.org/discover/newsletters/chronicle-winter-cport-2011.pdf.
23. City of Cambridge (2013).
24. Hotels.com (2015).
25. Xeroscleaning (2014).
26. United Nations Environmental Programme. (2005). *The Stakeholder Engagement Manual: Practitioner's Handbook on Stakeholder Engagement*. Vol. 2. Nairobi: United Nations Environmental Programme. Retrieved from http://www.accountability.org/images/content/2/0/208.pdf.
27. Accor Group (2014).
28. International Organization for Public Participation. (2015). Quality assurance standard for community & stakeholder engagement. Retrieved from https://www.iap2.org.au/coresoftcloud001/ccms.r?PageID=10122&tenid=IAP2.
29. Accor Group (2014).
30. Porter, M., and Kramer, M. (2011). Creating shared value. *Harvard Business Review*. Retrieved from https://hbr.org/2011/01/the-big-idea-creating-shared-value.
31. Taleb, N. (2007). *The Black Swan: The Impact of the Highly Improbable*. London, U.K.: Penguin Books.
32. Nolis, G. (2012). Laundry presentation. Hyatt Regency, Mumbai. Retrieved February 26, 2016, from http://www.sildeshare.net/grgnolis/laundry-presentation.
33. ISO (International Organization for Standardization). (2010). Guidance on use of the process approach. Geneva: ISO. Retrieved February 25, 2016, from http://www.iso.org/iso/04_concept_and_use_of_the_process_approach_for_management_systems.pdf.
34. Ibid.
35. U.S. Environmental Protection Agency. (2012). Saving water in hotels. Retrieved May 19, 2015, from https://www3.epa.gov/watersense/commercial/docs/factsheets/hotels_fact_sheet_508.pdf.

36. Xeros Cleaning. (2014). Near waterless laundry operations. Retrieved March 14, 2015, from http://www.Xeroscleaning.com/blog/emerging-trend-hotels-adopt-near-waterless-laundry-operations.

37. Jaffe, E. (2014). Read about how hotels get you to reuse towels: Everyone's doing it. Retrieved March 13, 2015, from http://www.fastcodesign.com/3037679?evidence/read-about-how-hotels-get-you-to-reuse-towels-every-oines -doing-it.

38. McChesney, C. (2012). Preview of the four disciplines of execution. Retrieved March 13, 2015, from http://www.franklincovey.com/4dflv/4D_bottomvid.html.

39. American Hotel and Lodging Association. (n.d.). Linen reuse program. Guideline 5. Retrieved May 9, 2015, from http://www.ahla.com/green.aspx?id=24954.

Appendix III: Olive Hotel Case—Part 2

This appendix presents the "check" and "act" elements of a virtual case developed by Cherie Mohr in a course that was taught using the method presented in this book. The information presented was completed before the final draft of this book was completed. Because this book is subtitled "Practical Step-by-Step Guide," this appendix can be used by the reader to develop some skills using the information presented in Chapters 15 through 19, as supplemented by the information in the foundation section, Chapters 1 through 6.

Skill building takes place in two steps. The first step involves understanding the information presented using the concepts outlined in the book. Once this has been completed, the reader should determine if each of the "essential questions" at the end of Chapters 15 through 19 is addressed in the case. You can also use Chapter 20 in the practice since it presents the introduction of the topic of organizational resilience that can be added to an organizational sustainability program.

While it is important to have knowledge about the elements of a sustainability program, it is critical to your success to develop the skills for using this knowledge for an organization that you are working with. Once you have reviewed this case, you should have developed the competence to apply the Section III information found in the book to an actual organization or another case. Unfortunately, there are very few cases available at this time that use this method for planning, implementing, and maintaining a fully embedded organizational sustainability program. More information may be found in the standards and reports described in Appendix I.

Scoping the Monitoring and Measurement Process (Chapter 15)

To create a basis for competitive advantage, the Olive Hotel must analyze the supporting elements in the value stream model.[1] These operations are provided in Figure 12.4. Together, these human functions help leadership think strategically about their operations. This kind of strategic thinking is necessary in order to identify, or scope, the significant activities the hotel needs to monitor to help meet its strategic objectives. The value chain model reveals linkages within the internal context, which will help the hotel optimize its internal value. There are linkages between the operations in the bottom half of the model that will help the organization seek opportunities for shared value[2] with Cambridge's Cambridgeport neighborhood.

Using the *lower portion* of the value chain model in Figure 12.4, the hotel should identify its primary activities involved in the creation of the services by which the hotel meets it objectives. These activities are related to operational processes and should be monitored for the purpose of maintaining effective processes and resource-efficient operations. Primary activities can be monitored using either past performance data (*lag* measures) or future performance drivers (*lead* indicators). Primary activities for the Olive Hotel involve suppliers, inputs of supplies, and people; processes that comprise the physical operations required to provide hotel services; outputs of services; and guest relations, including sales and marketing activities.

Using the *upper portion* of the model, the hotel should identify its support activities, which are the decision-making activities that support the primary activities. These activities are related to people and their performance, and are monitored for their effectiveness in executing operations to responsibly meet the hotel's strategic objectives over the long term. Support activities should be monitored using performance drivers (lead indicators), which are discussed further in Chapters 16 and 17. These support activities for the hotel include leadership and its management and strategic planning functions, human resources management in the hotel's internal context, information and knowledge management, stakeholder engagement and community development in the hotel's external context, and operations and procurement functions that keep the hotel facilities running efficiently. It is through these activities that the hotel delivers its services while managing opportunities and threats in its internal and external context, with the aim of creating shared value and competitive advantage within the community.[3] The following example illustrates how the Olive Hotel would use the value chain model to scope the opportunities for creating shared value with its Cambridgeport neighbors.

The hotel would start by using the value chain model to define the human resources (internal) and community development (external) activities, as shown in the "Organizational Support Operations" section in Appendix II. The hotel would then identify linkages between these activities, and explore how the human and process interactions within these linkages might pose uncertainties. Management could then design treatments for these uncertainties that would create shared value with neighbors and competitive advantage for the hotel. To illustrate using the Guest Partnership Program described in the above-mentioned section, the linkage between the external request for the hotel to measure and save water, and the internal improvement of the program created human and process interactions that posed uncertainty in the hotel's ability to achieve its objective of saving water. The uncertainty was rooted in the link between staff execution of program procedures designed to increase guest participation and staff perception that increased guest participation could destabilize their jobs if there is less laundry to do (Chapter 2). The hotel successfully treated this uncertainty by engaging internal stakeholders (staff) in goal-setting activities designed

Appendix III 283

to drive performance toward meeting the objective of saving water. The details of how this was accomplished are explained in the "Organizational Support Operations" (Appendix I) and "Measurement, Transparency, and Accountability" (this appendix) sections.

Scoping using the value chain model must be done periodically, and any time there is a change in the internal or external context, so the hotel can effectively manage its opportunities and threats. Together with uncertainty assessment, value chain scoping identifies the significant operations that help the hotel meet its objectives in an uncertain world, and which it should monitor in order to meet its environmental, social, and economic responsibilities over time. Proper scoping of the significant activities, which need to be monitored in order for the hotel to meet its objectives, is an essential part of an efficacious monitoring and measurement process, which will be discussed in the next section.

Monitoring and Sense Making (Chapter 16)

After the Olive Hotel has scoped its significant activities using a value chain model, it is ready to start monitoring its activities. *Monitoring* is the process for tracking progress toward meeting strategic and operational objectives. The purpose of monitoring and measurement is to improve the effectiveness and efficiency of operations. Like all operational processes, the monitoring and measurement process must have the necessary resources, defined methods, and steps for improvement and innovation. "What" to monitor and measure has been established by the *uncertainty assessment, stakeholder engagement*, and *value chain scoping* processes and by institutional and regulatory requirements. The monitoring and measurement processes address the "how."

Monitoring determines the status of operational performance compared with assumptions and expectations. In the hotel laundry, each of the 12 activities could be monitored to see how it is being performed. Monitoring is done by checking, supervising, and critically observing processes and their activities routinely and over time. The purpose is to ensure that the laundry operations' processes are effective and efficient, to detect (and ultimately predict) changes in both the internal and external context that might identify emerging opportunities and threats, and to gather information that can improve uncertainty analysis and sustainability. By monitoring the washing and drying activities, for example, hotel management might observe that the doors in the machines are significantly higher than the laundry carts, requiring the people loading and unloading them to repeatedly bend over. *Reactive* monitoring is used to detect deteriorating or failing processes; *proactive* monitoring is used to determine whether processes are operating as intended and objectives are being met. Both types of monitoring are necessary, but proactive monitoring catches irregularities before they turn into failures, thus reducing the need for reactive monitoring.

Monitoring the Washer/Dryer Operations

The monitoring process must include (1) the intended purpose; (2) which characteristics of the activity to monitor; (3) the method, metrics, equipment, timing, and controls; and (4) review and improvement of the process. The monitoring process should produce accurate and appropriate information, so as not to mislead or waste resources. Having bad data is worse than having no data at all. Applying these procedures, the process for monitoring the washing and drying activities might look like this:

1. The purpose of monitoring the washer/dryer (W/D) activities is to ensure the efficiency and safety of these activities.
2. Monitor W/D loading and unloading.
3. Management will visually observe the loading and unloading of carts and machines; record counts from the machines' digital load counters at the beginning and end of each individual's shift; measure and record the heights of the carts, machine doors, and individual operators; monitor all activities at all times for one month; and try to control the effects of observation on employee behavior (e.g., let employees know that everything is going well and that management will be observing various parts of the operation for improvement).
4. Management will record all observations in a log that includes date, time, and operator name; record all load counts in an MS Excel spreadsheet that includes date, time, and operator name.
5. Review data for correlations between obvious discomfort on the part of the operators and their corresponding load counts. (For example, are they resting between bending-over actions? Are they rubbing or supporting their backs, arms, legs, or necks? Does observable discomfort correlate with a given individual's productivity over time? Does the height of the operator correspond to observable discomfort and productivity?) It is important to stay focused on the purpose stated in number one above, which is to monitor the efficiency and safety of the activities, not to evaluate the productivity of individual operators. All correlations should be compiled into a report and presented to the hotel manager.

These monitoring procedures will enable management to determine whether the efficiency and safety of the washing and drying activities need to be improved and the most efficient and effective ways to improve them. It is leadership's responsibility to allocate adequate resources and expertise to monitor all hotel processes to ensure that they are effective and efficient. By doing so, the hotel will be able to manage opportunities and threats effectively and gain a strategic understanding of operational performance. It is easy to see how monitoring the washing and drying activities in the laundry will help leadership gain a strategic understanding of operational performance: the repeated bending over by machine operators poses a significant threat to job ergonomic safety, while the acquisition of taller carts poses a significant opportunity to increase productivity.

Monitoring Water Usage in the Hotel Laundry Operations

While the Olive Hotel's new parent organization, Accor Group, requires its hotels to monitor their laundry water usage,[4] the hotel has no established monitoring process for meeting this requirement. The hotel needs to first create a monitoring process. It can use the following approach to create a new monitoring program tailored to its context, which can be used for monitoring not only water usage in the laundry, but also any activity in the hotel's operations.

Using a plan–do–check–act model creates procedures for Olive Hotel's new monitoring program.

Plan

1. Establish objectives for the monitoring program that support the strategic objectives of the hotel.
2. Define the "target" process to be monitored.
3. Identify baseline measures and benchmark data from comparable processes or industry standards, with timelines, against which to compare progress.
4. Identify all activities within the target process.
5. Identify which activities within the process will be monitored.
6. Identify responsible persons associated with each monitoring activity.
7. Create appropriate metrics (lead and lag) for monitoring the process activities.
8. Define which metrics will be used to monitor each activity.
9. Identify and allocate the necessary resources needed to support the monitoring activities.
10. Define methods, schedules, and timelines for tracking, collecting, recording, compiling, and reporting data.
11. Create procedures for evaluating data for validity and relevancy, and compliance with the selected metrics.

12. Establish processes for analysis, improvement, and innovation.

Do

Monitor and record baseline measures. Put in place the necessary resources. Execute the planned monitoring activities. Collect, record, compile, and report data.

Check

Evaluate data for validity, relevancy, and compliance. Evaluate monitoring resources used against projections. Analyze the efficacy of the monitoring program for sense making and anticipating opportunities and threats in the hotel's operating environment.

Act

Modify the monitoring program to improve effectiveness and efficiency, uncertainty analysis inputs, and stakeholder engagement as required for transparency and sustainability. Explore opportunities for innovation and competitive advantage in the program based on lessons learned.

Using the above monitoring program procedures, the Olive Hotel can now monitor water usage in the hotel laundry operations. In order to quantifiably monitor the laundry water usage, the hotel will have to install submeters to isolate the laundry from the total hotel water usage. Meter readings of both laundry and total hotel water usage will provide lag measurements, which can be correlated with hotel occupancy figures corresponding to the scheduled meter readings. These meter readings and occupancy figures must be tracked, collected, recorded, compiled, and reported over time and according to an established schedule, to provide baseline data. This baseline data creates the lag indicators that will be used for evaluating the water usage results after action plans using lead indicators are implemented to modify laundry operations. As previously illustrated in the "Organizational Support Operations" section, reducing water usage in the hotel's laundry operations might be achieved by, among other things, monitoring one of the activities in the Guest Partnership Program to see how it could become more effective. The approach, deployment, assessment, and refinement (ADAR) method was used in the above-mentioned section to create a lead indicator to guide the level of performance that was sought in the activities. These activities in the Guest Partnership Program include messaging, laundering, guest room housekeeping procedures, monitoring, measurement, and record keeping. The program activity that was monitored was the guest room housekeeping procedures, and the lead indicator was increasing guest participation. An action plan, which defined the actions necessary to execute the housekeeping procedures, was developed to achieve the lead indicator. Housekeeping staff were then engaged in setting two work goals: reducing the number of soiled linen items per guest-night and pleasing the guests. Note that new monitoring procedures can be used to quantify improvements in guest participation; the staff can, if they choose, take their goal setting a step further by creating a contest for how to handle "keeper" towels so guests do not have to guess which dirty towel is theirs after housekeeping has hung them all up. The example was given of providing various colored clips for guests to attach to their towels, but the staff should engage in brainstorming other ideas. A floor-by-floor contest could energize staff by giving them recognition for their contributions in meeting the hotel's strategic objectives. This is an example of how a lead indicator drives performance by its clear rationale and defined purpose, focus on stakeholder interests, engagement of staff and guests, support of hotel objectives, and being measurable, innovative, and sustainable.

The monitoring and measurement process described in this section can provide the hotel with information about not only past performance and activities that drive future performance, but also the strength of the processes over time. This is called program "maturity" and will be discussed in the "Self-Assessment and Maturity" section. All three types of information are necessary to help the hotel create effective processes and efficient operations that will help it meet its objectives. There should be only one monitoring and measurement process, with a consistent set of metrics tailored to the hotel's principles and culture, and explicitly stated to make goals concrete and easily understood. This process can then be used on all activities at all levels of operation to ensure that the strategic objectives cascade down through the hotel's unique set of operations to the goals and action plans at the lowest levels of operation.

Sense Making

Like all processes in an organization, the process by which the Olive Hotel will ultimately achieve its objectives of saving water will require numerous decisions. Leadership must make decisions when assessing uncertainty in the external operating environment created by a stakeholder request for measurements, when developing new capabilities for saving water to deal with that request, and when committing resources to execute program improvements to manage the uncertainty. By having open and transparent dialogue with stakeholders, the hotel leadership has to read external context cues from the request for water usage data and sense the need to examine possible opportunities for water conservation. Leadership has to gather industry data to generate knowledge that will help them decide how to satisfy the stakeholders' underlying concerns in a more effective and economical way than if the hotel complies with the actual requests. Effective decision making on the part of the hotel leadership is based on successful "sense making," or understanding the dynamics of the external context so that they can make better decisions and the hotel can adapt and thrive.

Sense making will be especially important to the hotel in times of greater uncertainty in the external context, in things ranging from fluctuations in the cost and availability of resources, to new neighbors and competitors. When rapid changes occur in the external context and formal uncertainty analysis takes too long, competent sense making within hotel leadership will enable the hotel to coordinate an effective an timely response. This will make the Olive Hotel more resilient over time.

Measurement, Transparency, and Accountability (Chapter 17)

Measurement helps the hotel determine the value associated with operational performance, which can be quantitative or qualitatively stated. The measurement process consists of resources and processes related to measurement. In addition to the five steps in the hotel's monitoring process described in the "Monitoring and Sense Making" section, each measurement should include a clear statement of whether it is intended to indicate past results (lag measure) or future actions (lead indicator) that will meet a specific objective or goal. The measurement process must ensure valid results that are reliable, reproducible, and traceable, as well as appropriate for the intended purpose. The Olive Hotel leadership has the responsibility to make decisions in the midst of uncertainty, and measurements provide not only a method to analyze options for reducing uncertainty, but also the basis for information the hotel will share within its engagement with the stakeholders.

Measuring Water Usage in the Hotel Laundry Operations

The monitoring examples given in the "Monitoring and Sense Making" section, of using both lag and lead measures for monitoring water usage and reduction, illustrate how the specific activities being monitored determine the appropriate metrics, rather than the activities being determined by *preimposed metrics*. Lag measurements (meter readings) were needed to establish baselines and track and trend water usage for evaluating the effectiveness of water-saving activities and other changes in the laundry operations over time. A lead indicator (increased guest participation) was needed to drive performance toward meeting a goal. Meter readings (lag) can be conducted by anyone, without being tied to explicit goals or higher objectives. They are tied to organizational objectives only if they are conducted within an established monitoring and measurement program and become part of the overall data picture of the hotel. By contrast, increased guest participation (lead) is derived from organizational objectives, drives the performance of the employees executing the activities, and is therefore inherently embedded in the action plans for those activities.[5] Those activities, in turn, get adapted to meet the goals the employees have set, and thereby get tied back to the performance indicator. This is how lead indicators work from the bottom up to link goals at the lowest levels of operations to the strategic objectives at the top. The hotel should therefore seek to embed its metrics into its operations by primarily using lead indicators that support its overall operating strategy and, if feasible, lag measurements that benchmark and verify the results that meet its objectives. Sharing this strategy with stakeholders engages them in discussion about the drivers of productivity at all levels of the hotel, which not only satisfies the needs for transparency and accountability, but also creates shared value[6] that becomes the true stake in "stakeholder" and takes the Olive Hotel's sustainability program to a higher level.

Establishing Transparency and Accountability in Monitoring and Metrics

In the "Monitoring and Sense Making" section, an approach was presented for creating a monitoring and measurement program for monitoring anything within the hotel's operations. As pointed out in Chapter 15, anything can be measured. Even "intangibles" can be measured by defining them in terms of observable consequences and the decisions the measurements are supposed to support.[7] In order to be efficacious, the hotel's monitoring program must not only gather information from the internal and external context, but also share information with the context through the stakeholder engagement process. Transparency and accountability are essential in this sharing process. The Olive Hotel's parent organization, the Accor Group (internal stakeholder), and its customers and community (external stakeholders) sometimes request (lag) performance data from the Olive Hotel based on the perceived impact of the hotel's activities on their environmental, social, and economic interests. The hotel should strive to be transparent and accountable by addressing its stakeholders' interests in a clear, accurate, timely, honest, and complete manner; this transparency and accountability will help to create the trust, legitimacy, credibility, and respect necessary to obtain its social license to operate.[8] The hotel must also evaluate each request in terms of the required resources, reason for the request, significance of the perceived impact, and relevance to the hotel's objectives in order to determine how it will be accountable. By engaging stakeholders in transparent dialogue about their mutual interests regarding accountability, the hotel can influence the measures and metrics and align such requests with its own operations and objectives, all while enriching stakeholder interactions and maintaining its social license to operate.

Stakeholder Request for Hotel Landscape Water Usage

The Olive Hotel has received a specific request from an external stakeholder to provide a key performance indicator (KPI) for the amount of water it uses for outdoor landscaping, as a percentage of total water used by the hotel. The Olive Hotel understands the stakeholder request because hotels are very large consumers of water and account for about 15% of the total water used in commercial and industrial facilities, the second highest of all building types.[9] Most of the industry data available for hotel water usage is given in water meter measurements (lag indicators) expressed relative to total facility usage, as is the case for the KPI requested by the stakeholder. These indicators are very useful for giving a snapshot of the industry, and creating benchmarks and baselines for comparison.

The Olive Hotel, however, does not currently have a submeter installed for monitoring its outdoor landscape water. Monitoring landscape water is not required by Accor, and the hotel had not anticipated adding it to the monitoring program. In considering this request, therefore, the hotel should assess the reason the stakeholder might want the measurements, the resources required to install a submeter, the benefits to the hotel for having this measurement, and the significance of the potential impact of the hotel's landscape water usage on the stakeholder and on the environment.

The hotel management does not know why the stakeholder wants the measurements, but can nevertheless evaluate the request by looking at the measurement data from the perspective of the decisions the data is supposed to support.[10] From that perspective, the measurement value would lie in decisions regarding water conservation. The hotel manager can then reasonably assume that the stakeholder has more interest in the hotel reducing its landscape water usage. Adding a submeter to monitor landscape water usage is not in the budget, and moreover, installing a submeter and reading it will not save any water. The hotel should instead consider implementing performance (lead) indicators for saving water. When combined with action plans, these lead indicators will yield water savings, educate employees, and address the hotel's larger water conservation objectives. In terms of the significance of the impact of the hotel's landscape water usage on the stakeholder and the environment, data published by the Environmental Protection Agency (EPA) indicates[11] that the hotel's average water usage per function is as follows: guest rooms 30%, laundry 16%, landscaping 16%, food service 14%, heating and cooling 12%, swimming pools 1%, and other 12%. Average landscape water usage in hotels is significant, but the Olive Hotel may not fit the average profile due to its urban site and should focus instead on its much higher guest room water consumption and its existing plans to reduce its laundry water consumption.

A transparent conversation with the concerned stakeholder and the community about the hotel's various components of water usage is recommended. The hotel wishes to be accountable for its responsible use of resources and should therefore engage the external stakeholders over their shared interest in conserving water using the following approach:

1. Furnish the KPIs published by the EPA and make the observation that the Olive Hotel's KPI for landscape water usage is likely below 16% because it is on an urban site with very little landscaping and not turf. Since Cambridge, Massachusetts, is in a relatively verdant climate zone, the landscaping receives minimal irrigation.

2. Present a strategy for implementing lead performance indicators for saving landscape water. Action plans might include replacing spray irrigation heads with a drip system, planting native species, and installing rainwater collection devices at accessible downspouts.

3. Focus on the EPA data for hotel laundry water consumption, which, like landscaping water, has an average KPI of 16%. Make the observations that if the Olive Hotel's landscaping water KPI is less than 16%, then the hotel's other KPIs would be greater than the average data suggests, and so the hotel's laundry KPI would likely be greater than 16%.

4. Present the existing plan for using submetering and lead indicators to reduce water usage in the laundry operation through its Guest Partnership Program. Share industry data that suggests that an effective towel reuse program can reduce laundry loads by 17%.[12]

5. Present a phased plan to include conserving water in the guest rooms, which have an EPA average KPI of 30%, by adopting the EPA's Hotel Challenge and saving up to 20% in guest water usage by retrofitting guest bathrooms with ultralow water flow devices (e.g., low-flow flush toilets).[13]

The hotel depends on the external stakeholders for bringing up uncertainty in the external operating environment. Had it not been for this stakeholder request for landscape water usage data, the hotel may not have engaged in the scoping, monitoring, sense making, measurement, or transparent stakeholder engagement activities that ensued. This exercise illustrates how these sustainability practices can help the hotel become more resilient and maintain its social license to operate.

Self-Assessment and Maturity (Chapter 18)

Self-Assessment

In the previous section, we looked at how engaging stakeholders in transparent discussions about measuring water usage in the hotel landscaping operation led the Olive Hotel to create a program to actually save water, and not merely measure water usage. We looked at how the hotel managed the threat of expensive replumbing and installation of water metering equipment. We looked at how it used a leading indicator (i.e., improving guest participation in the Guest Partnership Program) to set employee goals, create action plans, and monitor and measure the results. Finally, we saw how effective stakeholder engagement, transparency, and accountability satisfied the stakeholders' interests and enabled the hotel to achieve its objective. This systems approach, together with the successful outcome, created shared value with the community. The creation of shared value helps the Olive Hotel maintain its social license to operate and takes its sustainability program to a higher level.

The hotel's progress along this path to sustainability is called "maturity." Maturity is evaluated using self-assessment, a comprehensive and systematic review of activities and performance using lead indicators to drive improvement, and it can be applied to any component of the hotel's sustainability program. Here, we look at how the hotel can use a self-assessment tool to assess its monitoring and measurement program (hereafter referred to as "program").

Elements

In the left column of Figure 18.2, we listed the elements ("subclauses" in the example) that characterize the key activities in the hotel's program. Use names that are tailored to the specific activities in the hotel's program, such as "Baselines and Benchmarks," "Mapping the Activities," "Metrics," "Methods," "Analysis and Risk," and "Improvement and Stakeholders."

Levels

The five levels across the top of the maturity matrix represent progress along the path to maturity from left (lowest) to right (highest). It is best to have a five-level model to provide easy milestones along the route to maturity of each element. Levels should be numbered for universal understanding of hierarchy and incremental differences; a one-word description can be added to each level if it adds to the clarity of the metric.

Each level in the maturity matrix is unique to the monitoring activities in the hotel's program and tailored to the hotel's context, operations, and the internal and external stakeholder's interests, and the lead indicators for achieving the sustainability objectives for each element. Taken as a whole, they should describe the full integration of the hotel's monitoring and measurement program with the other components of its sustainability program: strategic objectives, internal and external context, stakeholder engagement, governance, uncertainty assessment, efficient and effective operations, value chain scoping, transparency and accountability, maturity, and improvement and innovation. After the lead indicators in the far-right-side column are established, work backward to create descriptions for the other levels. You are essentially designing an action plan for each element where the actions are predictive and influenceable measures that lead to the goal on the far right—the sustainability objectives for that element.[14]

Like any action plan, each step along the way should be stated in terms of accomplishments and not deficiencies. Like any design process, this one is iterative and may require repeated rounds to complete. Creating the descriptions presents an opportunity to develop a "library" of terminology that will provide consistency and clarity when creating and using the self-assessment tool across all components of the sustainability program. This terminology will mature into a common language that employees and stakeholders will use to create top-down, bottom-up awareness of the hotel's objectives and embed environmental, social, and economic responsibility into every level of operations at the Olive Hotel.

Scores

Evaluate each element for maturity level based on the descriptions created. This provides a qualitative assessment of each element's progress along the path to maturity at the time of the evaluation. The output from the uncertainty assessment is six numeric scores, one for each of the six elements that make up processes and activities of the component being assessed.

Maturity

By tailoring the six elements of the Olive Hotel's monitoring and measurement program to the specific activities in the program—baselines and benchmarks, mapping the activities, metrics, methods, analysis and risk, and improvement and stakeholders—the numeric score for each of these six elements creates a snapshot of the maturity level for each element at a point in time. In order to track the maturity progress of each element over time, self-assessments should be conducted periodically according to a planned schedule. Each snapshot produces a unit of data that must be collected, recorded, complied with, and ultimately reported so it can be used by leadership for decision making. These data management activities will be defined by action plans in the hotel's new monitoring and measurement program.

To illustrate, let us assume that the hotel leadership wants to see the maturity progress of the improvement and innovation elements of its monitoring and measurement program. Periodic self-assessments would have been conducted using a self-assessment tool, as illustrated above. The self-assessments would have been accomplished according to an established schedule and timeline (e.g., every six months for two years), and the scores collected and recorded by a designated persons in a manner specified (e.g., an MS Excel spreadsheet) and saved in a designated place (e.g., the hotel's network server). When the saved data is requested, designated persons will retrieve the data, compile it into the format specified by the monitoring program for individual element maturity scores (e.g., graph), and give it to leadership. Leadership will prepare a letter or report that might include the data graphic, an explanation of the data and why it is important, and for what purpose it is being reported. Leadership would distribute the report to the appropriate parties for discussion of improvements and innovations in the improvement and innovation process in the monitoring and measurement program. The hotel should share this procedure with the Accor Group, which seems to be in need of improving its monitoring and reporting of laundry water consumption data for its environmental footprint life cycle analysis.[15]

The hotel will want to similarly track and report maturity progress for entire components, such as its monitoring and measurement program. The hotel's new monitoring program does not yet specify a format for communicating maturity snapshots of whole components. MS Excel radar or spider plots would be a good choice for expressing maturity data because they can convey multiple values (the element scores) over multiple common variables (the elements).[16] On the spider plot, connecting the points that represent the element scores forms a polygon. When comparing multiple maturity spider plots for the same component over time, the relative levels of maturity are immediately obvious by comparing the areas captured within the polygons. The polygons will change shape as different elements mature at different rates; however, the areas inside the polygons tell a clear story of the progress over time. This approach, therefore, would be particularly effective for showing the maturing of the hotel's monitoring and measurement program to stakeholders. The timeline of spider plots can be consolidated into a report and used to communicate to internal and external stakeholders the results of the hotel's efforts to improve the sustainability program. In addition to tracking maturity over time, leadership will use spider plots to compare maturity at a point in time among components in its sustainability program.

Improvement, Innovation, and Learning

All of the activities presented in this case are required to create continual improvement in the operations of the Olive Hotel. It must be kept in mind, that continual improvement is required to help the goals meet the objectives set by the hotel using the methods described in Chapter 7.

Innovation is used to make sure that the strategic objectives can be met over the long term. It is not sufficient to rely on continual improvement to make the necessary gains for the hotel to meet its strategic objectives. The

Accor Hotel Group must take a lead in driving innovation and sharing the best practices with the hotels that are affiliated with its brand.

Sustainability succeeds when there is a focus on learning at all levels of the facilities that constitute the corporation. There needs to be a constant focus on how best to share information and success throughout the corporate structure.

References

1. Porter, M.E. (1985). *Competitive Advantage: Creating and Sustaining Superior Performance*. New York: Free Press.
2. Porter, M. and Kramer, M.R. (2011). Creating shared value. *Harvard Business Review*. Retrieved from https://hbr.org/2011/01/the-big-idea-creating-shared-value.
3. Porter, M. and Kramer, M.R. (2011).
4. Accor Group. (2012). PLANET 21 Program. Retrieved March 13, 2015, from http://www.accorhotels-group.com/en/sustainable-development/planet-21-research.html
5. McChesney, C., Covey, S., and Huling, J. (2012). *The 4 Disciplines of Execution*. London: Simon & Schuster.
6. Porter, M. and Kramer, M.R. (2011).
7. Hubbard, D.W. (2010). *How to Measure Anything: Finding the Value of "Intangibles" in Business*. 2nd ed. Hoboken, NJ: John Wiley & Sons.
8. ISO (International Organization for Standardization). (2010). Social responsibility guidance. ISO 26000, Geneva: ISO.
9. US EPA. (2012b). *Water Use Tracking*. Retrieved from http://www.energystar.gov/ia/business/downloads/datatrends/DataTrends_Water_20121002.pdf?2003-40fb.
10. Hubbard, D.W. (2010).
11. US EPA (2012b).
12. Bruns-Smith, A. et al. (2015). Environmental sustainability in the hospitality industry. Retrieved from http://scholarship.sha.cornell.edu/cgi/viewcontent.cgi?article=1199&context=chrpubs.
13. US EPA (2012b).
14. McChesney, C., Covey, S., and Huling, J. (2012).
15. Bleu Safran. (2011). Beyond green travel: Critical review report. Retrieved from http://www.accor.com/fileadmin/user_upload/Contenus_Accor/Developpement_Durable/pdf/earth_guest_research/EGR_Empreinte_env/critical_review_report.pdf.
16. Excel Dashboard. (2013). Remove the zero in a radar chart. http://www.excel-dashboardtemplates.com/remove-the-zero-point-or-make-a-hole-in-an-excel-radar-chart/.

Index